Absorber Types in Vapour Absorption Refrigeration Systems

This book addresses the increasing energy demand and costs associated with the global refrigeration industry, primarily driven by the need for cooling. It proposes the substitution of vapour compression refrigeration systems (VCRS) with vapour absorption refrigeration systems (VARS), which operate on low-grade, renewable energy sources like solar, geothermal, and waste heat.

Focusing on the absorber component of VARS, which plays a critical role in facilitating heat and mass transfer processes, the book provides a comprehensive overview of absorber configurations, including tray, packed bed, falling film, spray, bubble, and membrane absorbers. It offers guidance on selecting the appropriate absorber configuration considering their advantages and limitations in different operating conditions, as well as their numerical, experimental, and performance enhancement studies.

The book will interest heating, ventilation, and air conditioning (HVAC) academic researchers, graduate students, and professionals involved in the advancement of sustainable refrigeration technologies, particularly absorber selection.

A. Mani is a professor in the Refrigeration and Air-Conditioning Laboratory in the Department of Mechanical Engineering at the Indian Institute of Technology Madras. He obtained his PhD from IIT Madras in 1986. He has undertaken more than 30 sponsored projects and consulting assignments from government organizations and private Ltd. companies. He has been the principal investigator for a number of projects sponsored by various agencies in India and abroad. He has been invited to many academic institutes around the globe for keynote and invited lectures. Professor Mani has published more than 218 research articles in reputed international journals and conferences. He is a reviewer for several international journals. He is a fellow of the Institution of Engineers (India). He is a life member of the Indian Society of Heating, Refrigeration and Air Conditioning Engineers (ISHRAE), Solar Energy Society of India, Indian Desalination Association (InDA), and Indian Society for Heat and Mass Transfer (ISHMT). He is a member of the editorial board of the *International Journal of Energy and Environment*.

Narasimha Reddy Sanikommu is a doctoral research scholar from the Refrigeration and Air Conditioning Laboratory, Department of Mechanical Engineering at the Indian Institute of Technology Madras. His research areas are Vapour Absorption Refrigeration Systems, HVAC, solar energy applications, experimental techniques, and multiphase flow simulations. He has presented his research at national and international conferences and published research articles in reputable international journals.

Absorber Types in Vapour Absorption Refrigeration Systems

A. Mani and Narasimha Reddy Sanikommu

CRC Press is an imprint of the
Taylor & Francis Group, an **informa** business

First edition published 2025
by CRC Press
2385 NW Executive Center Drive, Suite 320, Boca Raton FL 33431

and by CRC Press
4 Park Square, Milton Park, Abingdon, Oxon, OX14 4RN

CRC Press is an imprint of Taylor & Francis Group, LLC

ISBN: 978-1-032-77878-5 (hbk)
ISBN: 978-1-032-77879-2 (pbk)
ISBN: 978-1-003-48519-3 (ebk)

DOI: 10.1201/9781003485193

Typeset in Times
by KnowledgeWorks Global Ltd.

Contents

Preface

Over the past few decades, civilization has been experiencing high energy demand and high energy unit costs, mainly driven by increased cooling needs. The global refrigeration industry predominantly relies on the vapour compression refrigeration system (VCRS), which utilizes high-grade energy, a non-renewable energy source. One solution to mitigate these issues is substituting VCRS with the vapour absorption refrigeration system (VARS). Unlike VCRS, VARS operates on low-grade energy sources such as solar and geothermal, which are renewable.

Among all the components of VARS, an absorber is a critical component, as it facilitates two-component, coupled heat and mass transfer processes. Absorber plays a crucial role in determining the size and performance of the system under given operating conditions. This book provides an overview of various absorber configurations in the literature, including tray, packed bed, falling film, spray, bubble, and membrane absorbers. Also, numerical, experimental, and performance enhancement studies of each absorber are discussed separately. Finally, a comparison of absorbers is presented based on the findings of this work.

Each of the absorber configurations has its own advantages and disadvantages, and thus, careful consideration must be given to the selection of the appropriate absorber. For instance, falling film and plate-type absorbers are prone to wettability issues and film instabilities. On the other hand, spray and packed bed absorbers face difficulties with effective heat rejection. Bubble absorbers, however, possess a high interfacial area and good heat rejection capabilities. Membrane absorbers are best suited for low-capacity applications due to their lower operating temperatures. Based on a comparison of all the available absorber configurations, the bubble mode is the most suitable for low solution flow rates.

Nomenclature

ABBREVIATIONS

COP	coefficient of performance
d	diameter
DEGDME	di-ethylene glycol di-methyl ether
DMA	N, N-di-methyl acetamide
DMAC	dimethylacetamide
DMEDEG	dimethylether diethylene glycol
DMETEG	dimethyl ether of tetraethylene glycol
DMETrEG	dimethylether triethylene glycol
DMEU	dimethylol ethylene urea
DMF	N, N-di-methyl formamide
L	absorber length
$LiNO_3$	lithium nitrate
LMTD	logarithmic mean temperature difference
m	mass/volume flow rate
MCL	methylene chloride
NaSCN	sodium thiocyanate
NMP	N-methyl-2-pyrrolidone
P	absorber pressure
PEB6	pentaerythritol tetra-2-ethylbutanoate
PEB8	pentaerythritol tetra-2-ethyl hexanoate
PEC5	pentaerythritol tetra-pentanoate
PEC9	pentaerythritol tetra-nonanoate
PHE/PHX	plate heat exchanger
Re	Reynolds number
T	temperature
TEGDME	tetra-ethylene glycol di-methyl ether
TrEGDME	tri-ethylene glycol di-methyl ether
X	mass fraction

REFRIGERANTS

R1234yf	2, 3, 3, 3-tetrafluoropropene
R1234ze(E)	trans-1, 3, 3, 3-tetrafluoropropene
R124	1-chloro-1, 2, 2, 2-tetrafluoroethane
R125	pentafluoroethane
R134a	1, 1, 1, 2-tetrafluoroethane
R143a	1, 1, 1-trifluoroethane
R152a	1, 1-difluoroethane
R21	dichlorofluoromethane
R22	chlorodifluoromethane

R23	trifluoromethane
R32	difluoromethane

SUBSCRIPTS

c	coolant
c,in	coolant, in
o	orifice
s	solution
s,in	solution, in
s,out	solution, out
v	vapour
f	film

IONIC LIQUIDS

[BMIM][PF_6]	1-butyl-3-methylimidazolium hexafluorophosphate
[DMIM][Cl]	1, 3-dimethylimidazolium chloride
[DMIM][DMP]	1, 3-dimethylimidazolium dimethyl phosphate
[EMIM][AC]	ethyl-3-methylimidazolium acetate
[EMIM][BF_4]	l-ethyl-3-methylimidazoliumtetrafluoroborate
[EMIM][$EtSO_4$]	1-ethyl-3-methylimidazolium ethyl sulphate
[EMIM][OM]	1-ethyl-3-methylimidazolium methanesulphonate
[HMIM][BF_4]	1-hexyl-3-methylimidazolium tetrafluoroborate
[HMIM][Cl]	1-hexyl-3-methylimidazolium chloride
[HMIM][PF_6]	1-butyl-3-methylimidazolium hexafluorophosphat
[HMIM][Tf_2N]	1-hexyl-3-methylimidazolium bis (trifluoromethylsulphonyl) imide
[OMIM][BF_4]	1-octyl-3-methylimidazolium tetrafluoroborate
[OMIM][PF_6]	1-methyl-3-octylimidazolium hexafluorophosphate

1 Introduction

According to the Ministry of Food Processing Industries in India, the annual losses of primary agricultural products due to harvest and post-harvest processes are approximately INR 441.43 billion (at 2009 wholesale prices) and INR 926.51 billion (at 2014 wholesale prices) across the country (Ministry of Food Processing Industries, GOI, 2021–2022). Refrigerated cold storage can be an effective way to reduce these losses in different sectors. By maintaining a controlled temperature and humidity environment, cold storage can help extend the shelf life of perishable goods. This can help to reduce waste and improve the overall efficiency of the food supply chain. Refrigeration can also play a vital role in preserving the quality and safety of food, medicine, and other products, supporting industrial processes and shopping malls and providing comfort for individuals. The conventional method of refrigeration, the Vapour Compression Refrigeration System (VCRS), is the most common method of refrigeration used today. Running a VCRS requires high-grade energy, such as electrical energy. The Indian power sector heavily relies on coal as an energy source. The total power generation in India comprises 55% from coal, 21% from hydro, 11% from renewable energy sources, 10% from gas, and 3% from remaining sources (Bhattacharyya and Gupta, 2014). Also, the consumption of coal is more than the production, so the cost of electric energy is growing. If this coal-based electricity is used to run, a VCRS would likely result in higher running costs. However, it is worth noting that using coal-based electricity to run the VCRS can negatively impact the environment, including air pollution and greenhouse gas emissions. In recent years, there has been a push to diversify the power sector in India and incorporate more renewable energy sources, such as solar and wind, to reduce reliance on coal and mitigate these negative impacts. As a result, researchers have developed a Vapour Absorption Refrigeration System (VARS), which works based on low-grade renewable energy sources, such as solar energy, waste heat, and geothermal energy. These systems can be an attractive option in areas where access to electricity is limited or where traditional refrigeration systems, which rely on electricity or fossil fuels, are not desirable due to environmental concerns. One of the main benefits of VARSs is their ability to use low-grade energy sources, which are often abundant and widely available. This makes them a potentially cost-effective and sustainable alternative to traditional refrigeration systems, which can be expensive to operate and maintain. However, there are also some limitations to VARSs. They can be complex to design and install, requiring specialized maintenance and repair.

Additionally, the efficiency of VARSs can vary depending on the type of low-grade energy source being used, which may not be suitable for all applications. Overall, VARSs can be an attractive option for refrigeration in certain circumstances, particularly in areas with abundant low-grade energy sources and a need for efficient and sustainable refrigeration. With VARS, the greenhouse effect can be reduced using environmentally friendly refrigerants such as ammonia and water. VARS working with lithium bromide and water (LiBr–H_2O) requires a minimum solar energy

DOI: 10.1201/9781003485193-1

temperature of 80°C to give a cooling effect, which can be easily achieved using a simple solar flat plate collector (Kalogirou et al., 2001). Hence, VARS is considered to be an alternative to VCRS. VARS is used in supermarkets, hotels, and refrigeration industries for air conditioning and cooling applications.

1.1 VAPOUR ABSORPTION REFRIGERATION CYCLE OPERATING PRINCIPLE

Among the various refrigeration systems, VCRS and VARS are the most widely used systems. VARS, unlike VCRS, uses low-grade thermal energy. As previously stated, VARS is more effective than VCRS in places where low-grade energy is available. The main difference between VCRS and VARS is in how it compresses the refrigerant vapour. In VCRS, the compression process is achieved using a mechanical compressor; whereas in VARS, it is achieved using a thermal compressor. Schematic diagrams of VCRS and VARS are shown in Fig. 1.1.

Figure 1.1 shows that the mechanical compressor in VCRS is entirely replaced by the absorber, solution pump, generator, and pressure reduction valve in VARS. Interestingly, the high-grade energy required for the pump in VARS is comparatively negligible compared to that of the compressor in VCRS due to its liquid working fluid, and also, it is 2–3% of the generator heat load in VARS. It leads to the conclusion that the solution pump work is negligible compared to the heat input in VARS and the mechanical energy required in VCRS. However, the performance of both cycles cannot be compared based on the first law efficiency (coefficient of performance (COP)) since the energy input in both cycles is a different grade of energy. The best cycle should be determined by its second law efficiency, which is of a similar order in both cycles, according to many researchers. VCRS uses a single-component working fluid, while VARS uses a multi-component working fluid, so analysing VARS is more complex than VCRS.

The basic schematic and cycle diagrams of a single-stage, single-effect VARS with an internal solution heat exchanger (SHX) are shown in Figs. 1.2 and 1.3, respectively.

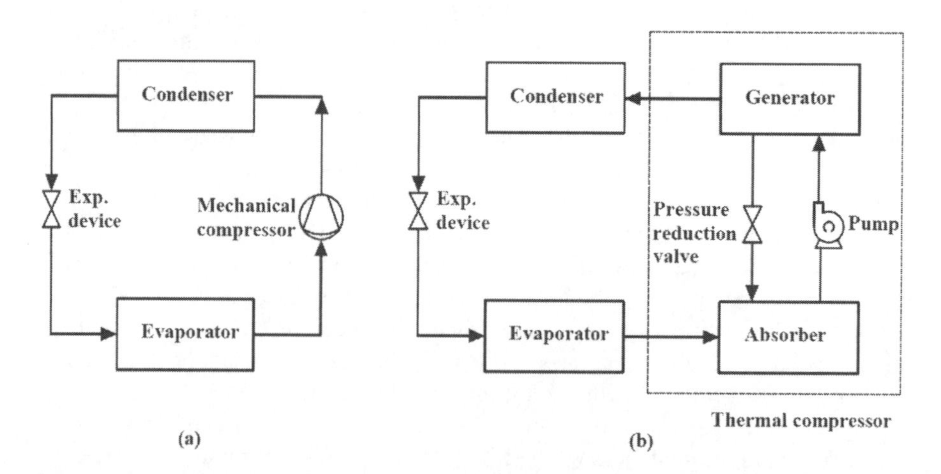

FIGURE 1.1 Basic schematic diagrams: (a) VCRS; (b) VARS.

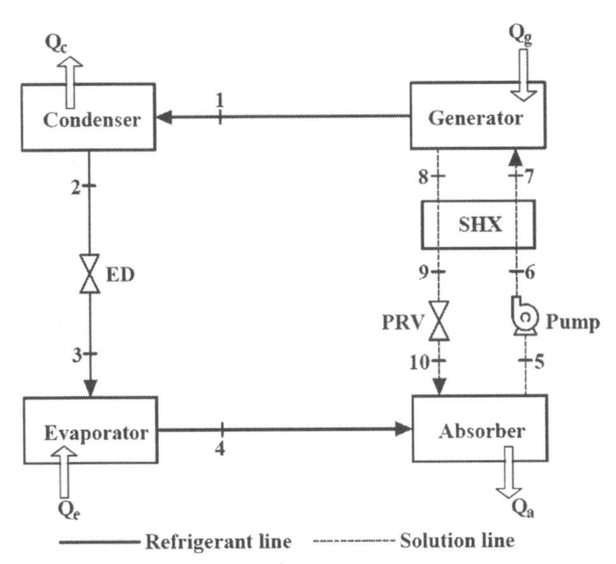

FIGURE 1.2 Schematic diagram of VARS with SHX.

The VARS operates by adopting all VCRS processes, including condensation, expansion, and evaporation. The compression process in VARS is performed by an absorber, pump, pressure reduction valve, and generator, known as a thermal compressor. Conventionally, in VARS, the refrigerant used is ammonia, water, etc. The refrigerant gets condensed in the condenser using either water cooling or air cooling, and then it

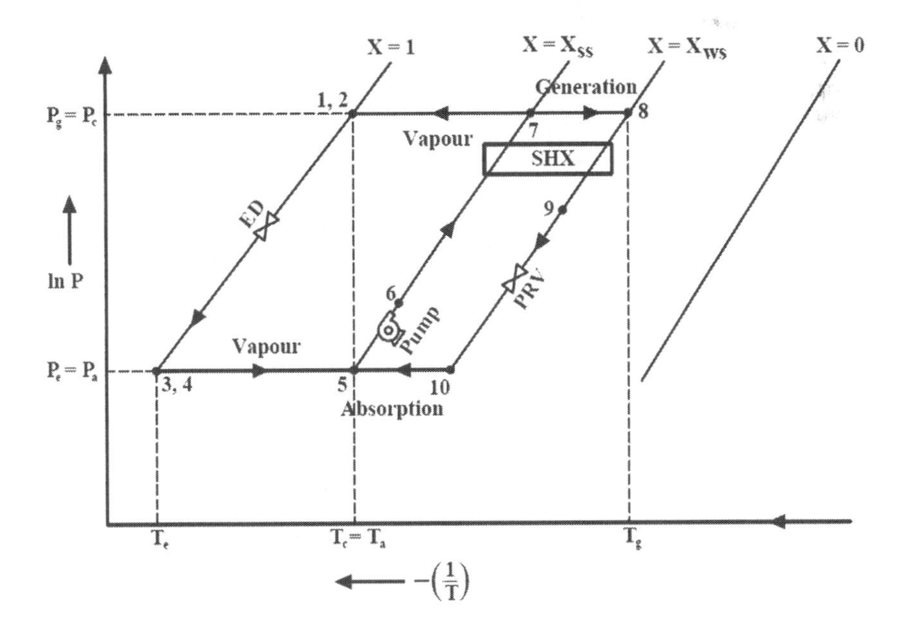

FIGURE 1.3 Cycle diagram of VARS with SHX.

expands to low pressure (evaporator pressure) through a throttle valve. This results in a refrigerant temperature lower than that of the product being cooled. The refrigerant evaporates in the evaporator by absorbing the heat from the product, thereby producing a refrigeration effect. The resulting refrigerant vapour is absorbed by the weak solution (less concentrated refrigerant solution) in the absorber, resulting in strong solution formation. In this process, because it is an exothermic reaction, the liberated heat must be removed using an external cooling media to maintain a constant temperature in the absorber and sustain the absorption process further. The pressure of this strong solution is increased by the solution pump to high pressure (generator pressure). Then, by adding low-grade heat to the strong solution, the refrigerant vapour is liberated in the generator due to the difference in boiling point temperature between the refrigerant and absorbent liquids. This refrigerant vapour is condensed in the condenser, and the remaining weak solution in the generator is sent back to the absorber through a pressure reduction valve. Thus, it completes the cycle. An internal SHX is an add-on component of the VARS to enhance performance. It exchanges the heat of weak solution with strong solution so that the amount of heat input at the generator and the amount of heat to be removed at the absorber gets reduced.

The present monograph discusses the performance parameters of VARS and the working fluids used in VARS in the following sections. Then, it examines various absorber configurations, namely, tray or plate type, packing bed, falling film, spray or adiabatic, bubble, and membrane absorbers, with emphasis on its numerical and experimental works. Finally, absorbers are compared based on their performance parameters.

1.2 PERFORMANCE PARAMETERS

The performance of VARS is evaluated by using various performance parameters. The most significant performance parameters of VARS are circulation ratio (CR), Carnot and thermodynamic COPs (CarCOP and COP), second law efficiency (IIeff), and the heat and mass transfer effectiveness of the heat exchangers such as absorber, generator, condenser, evaporator, and internal heat exchanger (Fatouh and Murthy, 1996a).

a. **Circulation ratio** is the amount of weak solution that needs to be circulated in the absorber to absorb 1 kg of refrigerant. It can be determined by performing a mass balance across the absorber as follows:

$$CR = \frac{1 - x_{ws}}{x_{ss} - x_{ws}} \quad (1)$$

It can also be defined as the ratio of the strong solution mass flow rate to the refrigerant mass flow rate.

$$CR = \frac{\dot{m}_{ss}}{\dot{m}_r} \quad (2)$$

b. **Carnot COP** is the maximum possible COP under given operating conditions. For a simple case where the absorber temperature is equivalent to the condenser temperature, it can be defined as

$$CarCOP = \left(\frac{T_g - T_a}{T_g}\right)\left(\frac{T_e}{T_c - T_e}\right) \quad (3)$$

c. **Thermodynamic COP** is defined as the ratio of cooling capacity to the power consumption, that is,

$$COP = \frac{Q_e}{Q_g + W_p} \quad (4)$$

However, the energy consumed by the pump is negligible compared to the energy supplied by the generator. Therefore, it can be simplified to

$$COP = \frac{Q_e}{Q_g} \quad (5)$$

d. **Second law efficiency** is defined as the ratio of the useful exergy at the evaporator to the exergy input at the generator. It is a measure of the extent to which the cycle is approaching ideal behaviour.

$$IIeff = \frac{COP}{CarCOP} \times 100 \quad (6)$$

e. **Effectiveness** of a heat exchanger is the ratio of actual heat transfer to the maximum possible heat transfer. Similarly, the mass transfer effectiveness for the absorber and generator is defined as

$$\epsilon_a = \frac{X_{ss} - X_{ws}}{X_{ss,eq} - X_{ws}} \text{ and } \epsilon_g = \frac{X_{ss} - X_{ws}}{X_{ss} - X_{ws,eq}} \quad (7)$$

Thermodynamic analysis of single-stage VARS working with R22-DMA, R22-DMF, and R22-DMETEG and the influence of operating temperatures on system performance is investigated by Fatouh and Murthy (1996b). The results showed that the COP, CR, and IIeff decrease as the generator temperature (T_g) decreases. Conversely, the CR and IIeff decrease, while the COP increases as condenser (T_c) and absorber (T_a) temperatures decrease. Lower evaporator temperatures (T_e) result in higher CR values but lower COP and IIeff values. The study concluded that R22-DMA is the best working pair for COP and IIeff. Another investigation examined the effects of VARS component effectiveness on a single-stage VARS working with R22-DMA, R22-DMF, and R22-DMETEG by Fatouh and Murthy (1996a). The results showed that high COP and IIeff and low CR are achieved with high ϵ_a, ϵ_h, and ϵ_g. The effect of

ϵ_h is more dominant in the system performance than that of ϵ_a and ϵ_g. The study also found that CR is affected more by ϵ_a than the ϵ_g and it is not dependent on ϵ_h.

A study is conducted to study the effect of operating temperatures on the performance of single-stage solar intermittent VARS using R134a-DMF, R32-DMF, and R22-DMF (He et al., 2009). The results indicated that as T_g increases, CR decreases, and COP increases up to a certain point, then COP decreases for all the working pairs studied. Among the three working pairs, R134a–DMF has shown the best performance. However, contrasting findings for R134a–DMF are observed in another study (Yokozeki, 2005). As T_g decreases, CR and COP increase, and as T_a decreases, CR decreases while COP increases. Another study shows that as T_e increases, COP increases while exergy efficiency decreases (Anand et al., 2018). Additionally, COP and exergy efficiency decrease as the absorber and condenser temperature increase.

Kaynakli and Kilic (2007) conducted studies on the performance of the LiBr–H_2O VARS under changing operating conditions. The authors found that as T_g and T_e increase, the COP increases continuously and IIeff increases up to some point and then decreases. Similarly, as T_c and T_a increase, COP decreases and IIeff follows the same trend as T_g and T_e. The high SHX effectiveness improves the system performance by 44%. Some working pairs are unable to perform better at sub-zero temperatures. To resolve this performance issue, Songara and Murthy (1998b) demonstrated the double effect VARS to increase COP and successfully achieve sub-zero temperatures using two single-stage VARS, even though the COP of R134a–DMA is less than R22-DMA. Other studies have also explored the performance of different working pairs, including NH_3–NaSCN (Tyagi, 1986), aqua ammonia (Tyagi, 1987), and exergy analysis of LiBr–H_2O for half effect, single effect, double effect, and triple effect systems (Maryami and Dehghan, 2017), etc. However, these advanced cycles increase system complexity and cost, affecting its feasibility, even though performance is enhanced.

The preceding work has focused on the parallel arrangement of cooling liquid to the absorber and condenser. However, it is possible to encounter a series of cooling water connections in reality. When utilizing a series connection, it is advantageous to send the cooling liquid to the condenser first and then to the absorber when high cooling liquid temperatures are available (Murthy, 1996). If the cooling liquid temperature is low, sending it to the absorber first and then to the condenser for R22-based VARS is recommended. The same analysis is also conducted for R21-DMF, R21-DMETEG, and R22-DMF by Kumar et al. (1991).

REFERENCES

Anand, Y., Tyagi, S.K., Anand, S., 2018. Variable capacity absorption cooling system performance for building application. J. Therm. Eng. 4, 2303–2317.

Bhattacharyya, B., Gupta, V.K., 2014. Present and future energy scenario in India. J. Inst. Eng. India Ser. B. 95, 247–254. https://doi.org/10.1007/s40031-014-0099-7

Fatouh, M., Murthy, S.S.A.S.A., 1996a. HCFC22-based vapour absorption refrigeration systems. Part II: Influence of component effectiveness. Int. J. Energy Res. 20, 371–384.

Fatouh, M., Murthy, S.S.A.S.A., 1996b. HCFC22-based vapour absorption refrigeration system: Part I: Parametric studies. Int. J. Energy Res. 20, 297–312.

He, L.J., Tang, L.M., Chen, G.M., 2009. Performance prediction of refrigerant-DMF solutions in a single-stage solar-powered absorption refrigeration system at low generating temperatures. Sol. Energy 83, 2029–2038. https://doi.org/10.1016/j.solener.2009.08.001

Kalogirou, S., Florides, G., Tassou, S., Wrobel, L., 2001. Design and construction of a lithium bromide water absorption refrigerator. CLIMA 2000/Napoli 2001 World Congr. 15–18.

Kaynakli, O., Kilic, M., 2007. Theoretical study on the effect of operating conditions on performance of absorption refrigeration system. Energy Convers. Manag. 48, 599–607. https://doi.org/10.1016/j.enconman.2006.06.005

Kumar, S., Prévost, M., Bugarel, R., 1991. Comparison of various working pairs for absorption refrigeration systems: Application of R21 and R22 as refrigerants. Int. J. Refrig. 14, 304–310. https://doi.org/10.1016/0140-7007(91)90046-J

Maryami, R., Dehghan, A.A., 2017. An exergy based comparative study between LiBr/water absorption refrigeration systems from half effect to triple effect. Appl. Therm. Eng. 124, 103–123. https://doi.org/10.1016/j.applthermaleng.2017.05.174

Ministry of Food Processing Industries, 2022. Annual Report 2021-22, Ministry of Agriculture & Farmers Welfare Government of India. https://www.mofpi.gov.in/sites/default/files/mofpi_annual_report_for_web_english.pdf

Murthy, S.S., 1996. HCFC22-based absorption cooling systems, part iii: Effects of different absorber and condenser temperatures. Int. J. Energy Res. 20, 483–494.

Songara, A.K., Murthy, S.S., 1998. Thermodynamic studies on HFC134a-DMA double effect and cascaded absorption refrigeration systems. Int. J. Energy Res. 614, 603–614.

Tyagi, K.P., 1986. Second law analysis of NH3-NaSCN absorption refrigeration cycle. Heat Recover. Syst. 6, 73–82.

Tyagi, K.P., 1987. Aqua-ammonia heat transformers. Heat Recover. Syst. CHP 7, 423–433. https://doi.org/10.1016/0890-4332(87)90004-4

Yokozeki, A., 2005. Theoretical performances of various refrigerant-absorbent pairs in a vapor-absorption refrigeration cycle by the use of equations of state. Appl. Energy 80, 383–399. https://doi.org/10.1016/j.apenergy.2004.04.011

2 Working Fluids Used in VARS

The chemical and thermodynamic properties of working fluids significantly influence the system's performance and initial cost. The conventional working fluids used in vapour absorption refrigeration system (VARS) are lithium bromide–water (LiBr–H_2O) and ammonia–water (NH_3–H_2O). In 1859, VARS working with ammonia–water is introduced and used to produce ice and store food. Later, in the 1950s, lithium bromide–water-based VARS is developed for industrial applications (Srikhirin et al., 2000; Kaynakli and Kilic, 2007). Conventional fluids have some disadvantages, such as crystallization, vacuum operating pressures with lithium bromide–water and rectification, and toxicity problems with ammonia–water. Because of these disadvantages with conventional fluids, research has been started to overcome these disadvantages and improve the system's performance, leading to the invention of alternative refrigerant–absorbent pairs.

The requirements of refrigerant–absorbent are as follows (Srikhirin et al., 2000; Abumandour et al., 2017; Merkel et al., 2018):

1. These are ready to mix with each other within a given range of operating temperatures, resulting in a high concentration of the absorbent solution. This helps to maintain low circulation ratios per unit cooling capacity.
2. Refrigerants should have high latent heat of vaporization to minimize the circulation ratio.
3. The transport properties of a solution should have high thermal conductivity, low viscosity, low specific heat, and low surface tension.
4. The refrigerant and absorbent should be non-corrosive, stable, non-toxic, non-explosive, and eco-friendly.
5. The solution should exhibit a large negative deviation from Raoult's law.
6. The heat of mixing should be minimal.
7. Large boiling point difference between refrigerant and absorbent is required.
8. The addition of absorbent should result in a notable reduction of the vapour pressure of the refrigerant.
9. These should be cost-effective and readily available.

The refrigerants are classified into four generations: First, second, third, and fourth. First-generation refrigerants are natural refrigerants such as carbon dioxide, ethers, and sulphur dioxide. The problems with first-generation refrigerants are flammability and toxicity. Second-generation refrigerants such as CFCs, HCFCs, ammonia, and water have evolved for safety, and ozone depletion is happening because of CFCs and HCFCs. So, the Montreal Protocol (1987) banned these refrigerants. HFCs, ammonia, water, etc., belong to the third-generation refrigerants as

DOI: 10.1201/9781003485193-2

non-ozone-depleting alternative refrigerants. However, these refrigerants have considerable global warming potential values. Hence, the Kyoto Protocol (1997) targeted banning greenhouse gas emission refrigerants. So fourth-generation refrigerants have been invented, and they belong to the refrigerant category, which gives low GWP, like HFO (Calm, 2008).

Ammonia–water and lithium bromide–water have been widely studied in the open literature. In a comparative study of P-T-x-h data for R22 refrigerant with six different absorbents (DMA, DMF, NMP, DMEDEG, DMETEG, and DMETrEG), it is found that R22 with DMA and NMP are preferred for VARS applications due to its lower heat of mixing (Fatouh and Murthy, 1993b). The same authors also developed a correlation for the minimum generator temperature required based on the evaporator, condenser, and absorber temperature for the same working fluids (Fatouh and Murthy, 1993a).

A thermodynamic analysis has been conducted on 26 refrigerant–absorbent combinations, which has led to the conclusion that 1,4-butanediol is the suitable absorbent for ammonia–organic solvent systems. The results have shown that the coefficient of performance (COP) increases as the percentage of salt added to ammonia-organic solvent systems increases, and sulphur dioxide with DMF and DMA are better solvents on a weight basis. Tyagi and Rao (1984) suggest that the R21-DMF is the better working fluid due to the minimum circulation ratio for this combination. Additionally, for R22, it has been found that DMF and DMA are better solvents (Agarwal and Bapat, 1985).

Experimental tests are conducted on R124 as an alternative refrigerant to CFCs and R22, using NMP, MCL, DMAC, DMEU, and DMETEG as absorbents. The results suggested that R124–DMAC is the best working pair among the aforementioned (Borde et al., 1997). P-T-x data and heat of mixing for R22 are presented in experiments using N-N-dimethyl acetamide, diethylene glycol, cyclohexanone, xylene and aniline, and R12-N, N dimethyl acetamide and R12-cyclohexanone. The experiments are conducted over a temperature range of 0°C–100°C and a concentration range of 0.05–0.95 (Bhaduri and Varma, 1985, 1986, 1988).

A model that is thermodynamically consistent and based on the equation of state for refrigerant–absorbent pairs is demonstrated by Yokozeki, 2005. The refrigerants used in the model are R22, R32, R125, R134, R134a, R143a, R152a, and ammonia and water. The absorbents used in the model are PEC5, PEC9, PEB6, PEB8, DMA, DMF, DEGDME, TrEGDME, TEGDME, and NMP. Among these working pairs, the circulation ratio is the least for R22-DMF, followed by R134a-DMF and R22-TEGDME. Lithium bromide–water is observed to have the highest COP, followed by ammonia–water and R22-DMF.

Songara and Murthy (1998a) suggested R134a as a potential alternative refrigerant for VARS. They conducted a performance analysis of R134a in combination with DMETEG, MCL, and DMEU as absorbents. Results showed that R134a has superior qualities to R22, including higher COP and lower circulation ratio. Moreover, Borde et al. (1995) also found that R134a–DMETEG is the best working pair compared to the other two working pairs (R134a–MCL and R134a–DMEU). Nezu et al. (2002) conducted a study on the suitability of R134a as a refrigerant with absorbents such as DMA, DMF, DEGDME, TrEGDME, and TEGDME. From the results, the authors developed a correlation for the bubble point pressure of all the working fluids using the static variable method and also developed enthalpy–concentration charts for the studied working pairs.

A study is conducted to compare the performance of different mixtures, namely R22-DMF, R134a-DMF, and R32-DMF, in a single-stage solar intermittent absorption refrigeration system. The results showed that R134a–DMF is the most suitable working fluid in the range of evaporator temperatures from 278K to 288K (He et al., 2009). Isothermal vapour-liquid equilibrium (VLE) data and solubility characteristics of R134a–DMF are given by Zehioua et al. (2010) and Han et al. (2011), respectively. Other researchers have also explored different alternative refrigerant–absorbent pairs like R23–DMF (Gao et al., 2012), NH_3–$LiNO_3$ and NH_3–NaSCN (Sun, 1998), R23+R134a-DMF (He et al., 2005), R134a+R32-DMF (Yin and Zhang, 2003), R1234yf-NMP, R22-DMA and R134a-DMA (Songara and Murthy, 1998), and R1234yf-DMETrEG (Fang et al., 2018), etc.

Ionic liquids (IL) have been investigated as potential absorbents for VARS. Ionic liquids offer several advantages, such as high absorption capacity, low vapour pressure, good thermal stability, and low volatility (Abumandour et al., 2017). A review of H_2O–IL mixtures, including H_2O-[DMIM][Cl], H_2O-[DMIM][DMP], H_2O-[EMIM][BF_4], and H_2O-[EMIM][$EtSO_4$] by Abumandour et al. (2017). Research in the field of ionic liquids is expanding rapidly. Some of the ionic liquids for VARS in the open literature are ammonia as a refrigerant with water as a co-solvent, and the absorbents are of [EMIM][$EtSO_4$], [EMIM][AC] and [HMIM][Cl] (Swarnkar et al., 2014), water-EMIM OM (Merkel et al., 2018), R1234ze(E) with [BMIM][PF_6], [HMIM][PF_6] and [OMIM][PF_6] (Sun et al., 2018), R1234ze(E) with [EMIM][BF_4], [HMIM][BF_4], [OMIM][BF_4] and [HMIM][Tf_2N] (Wu et al., 2018), and R134a-[HMIM][PF_6], [HMIM][Tf_2N] (Kim and Kohl, 2014).

REFERENCES

Abumandour, E.S., Mutelet, F., Alonso, D., 2017. Are ionic liquids suitable as new components in working mixtures for absorption heat transformers. Progress and Developments in Ionic Liquids Progress and Developments in Ionic Liquids; Handy, S., Ed.; InTech: Rijeka, Croatia, p.1. https://doi.org/10.5772/65756

Agarwal, R.S., Bapat, S.L., 1985. Solubility characteristics of R22-DMF refrigerant-absorbent combination. Int. J. Refrig. 8, 70–74. https://doi.org/10.1016/0140-7007(85)90076-3

Bhaduri, S.C., Varma, H.K., 1985. PTX behaviour of refrigerant-absorbent pairs. Int. J. Refrig. 8, 172–176. https://doi.org/10.1016/0140-7007(85)90158-6

Bhaduri, S.C., Varma, H.K., 1986. P-T-X behaviour of R22 with five different absorbents. Int. J. Refrig. 9, 362–366. https://doi.org/10.1016/0140-7007(86)90009-5

Bhaduri, S.C., Varma, H.K., 1988. Heat of mixing of R22-absorbent mixtures. Int. J. Refrig. 11, 92–95. https://doi.org/10.1016/0140-7007(88)90119-3

Borde, I., Jelinek, M., Daltrophe, N.C., 1995. Absorption system based on the refrigerant R134a. Int. J. Refrig. 18, 387–394. https://doi.org/10.1016/0140-7007(95)98161-D

Borde, I., Jelinek, M., Daltrophe, N., 1997. Working fluids for an absorption system based on R124 (2-chloro-1,1,1,2,-tetrafluoroethane) and organic absorbents. Int. J. Refrig. 20, 256–266. https://doi.org/10.1016/S0140-7007(97)00090-X

Calm, J.M., 2008. The next generation of refrigerants - historical review, considerations, and outlook. Int. J. Refrig. 31, 1123–1133. https://doi.org/10.1016/j.ijrefrig.2008.01.013

Fang, Y., Guan, W., Bao, K., Wang, Y., Han, X., Chen, G., 2018. Isothermal vapor-liquid equilibria of the absorption working pairs (R1234yf + NMP, R1234yf + DMETrEG) at temperatures from 293.15 K to 353.15 K. J. Chem. Eng. Data 63, 1212–1219. https://doi.org/10.1021/acs.jced.7b00821

Fatouh, M., Murthy, S.S., 1993a. Comparison of R22-absorbent pairs for absorption cooling based on P-T-X data. Renew. Energy 3, 31–37.

Fatouh, M., Murthy, S.S.A.S.A., 1993b. Comparison of R22-absorbent pairs for vapour absorption heat transformers based on P-T-X-h data. Heat Recover. Syst. CHP 13, 33–48.

Gao, Z., Xu, Y., Li, P., Cui, X., Han, X., Wang, Q., Chen, G., 2012. Solubility of refrigerant trifluoromethane in N,N-dimethyl formamide in the temperature range from 283.15 K to 363.15 K. Int. J. Refrig. 35, 1372–1376. https://doi.org/10.1016/j.ijrefrig.2012.03.002

Han, X., Gao, Z., Xu, Y., Qiu, Y., Min, X., Cui, X., Chen, G., 2011. Solubility of refrigerant 1,1,1,2-tetrafluoroethane in the N, N-dimethyl formamide in the temperature range from (263.15 to 363.15) K. J. Chem. Eng. Data 56, 1821–1826. https://doi.org/10.1021/je100975b

He, Y., Hong, R., Chen, G., 2005. Heat driven refrigeration cycle at low temperatures. Chinese Sci. Bull. 50, 485–489. https://doi.org/10.1007/BF02897466

He, L.J., Tang, L.M., Chen, G.M., 2009. Performance prediction of refrigerant-DMF solutions in a single-stage solar-powered absorption refrigeration system at low generating temperatures. Sol. Energy 83, 2029–2038. https://doi.org/10.1016/j.solener.2009.08.001

Kaynakli, O., Kilic, M., 2007. Theoretical study on the effect of operating conditions on performance of absorption refrigeration system. Energy Convers. Manag. 48, 599–607. https://doi.org/10.1016/j.enconman.2006.06.005

Kim, S., Kohl, P.A., 2014. Analysis of [hmim][PF6] and [hmim][Tf2N] ionic liquids as absorbents for an absorption refrigeration system. Int. J. Refrig. 48, 105–113. https://doi.org/10.1016/j.ijrefrig.2014.09.003

Merkel, N., Bücherl, M., Zimmermann, M., Wagner, V., Schaber, K., 2018. Operation of an absorption heat transformer using water/ionic liquid as working fluid. Appl. Therm. Eng. 131, 370–380. https://doi.org/10.1016/j.applthermaleng.2017.11.147

Nezu, Y., Hisada, N., Ishiyama, T., Watanabe, K., 2002. Thermodynamic properties of working-fluid pairs with R-134a for absorption refrigeration system. IIR/IIF-Commission B1, B2, E1 E2 446–453.

Songara, A.K., Murthy, S.S., 1998. Comparative performance of HFC134a- and HCFC22-based vapour absorption refrigeration systems. Int. J. Energy Res. 372, 363–372.

Srikhirin, P., Aphornratana, S., Chungpaibulpatana, S., 2000. A review of absorption refrigeration technologies. Renew. Sustain. Energy Rev. 5, 343–372. https://doi.org/10.1016/S1364-0321(01)00003-X

Sun, D.W., 1998. Comparison of the performances of NH3-H2O, NH3-LiNO3 and NH3-NaSCN absorption refrigeration systems. Energy Convers. Manag. 39, 357–368. https://doi.org/10.1016/s0196-8904(97)00027-7

Sun, Y., Zhang, Y., Di, G., Wang, X., Prausnitz, J.M., Jin, L., 2018. Vapor-liquid equilibria for R1234ze(E) and three imidazolium-based ionic liquids as working pairs in absorption-refrigeration cycle. J. Chem. Eng. Data 63, 3053–3060. https://doi.org/10.1021/acs.jced.8b00314

Swarnkar, S.K., Srinivasa Murthy, S., Gardas, R.L., Venkatarathnam, G., 2014. Performance of a vapour absorption refrigeration system operating with ionic liquid-ammonia combination with water as cosolvent. Appl. Therm. Eng. 72, 250–257. https://doi.org/10.1016/j.applthermaleng.2014.06.020

Tyagi, K.P., Rao, K.S., 1984. Choice of absorbent-refrigerant mixtures. Int. J. Energy Res. 8, 361–368. https://doi.org/10.1002/er.4440080406

Wu, W., You, T., Zhang, H., Li, X., 2018. Comparisons of different ionic liquids combined with trans-1,3,3,3-tetrafluoropropene (R1234ze(E)) as absorption working fluids. Int. J. Refrig. 88, 45–57. https://doi.org/10.1016/j.ijrefrig.2017.12.011

Yin, Y., Zhang, X., 2003. Experimental study on a new absorption refrigeration system using R134a+R32/DMF. In Proceedings of the 5th Minsk International Seminar "Heat Pipes, Heat Pumps, Refrigerators", Minsk, Belarus (pp. 315–319).

Yokozeki, A., 2005. Theoretical performances of various refrigerant-absorbent pairs in a vapor-absorption refrigeration cycle by the use of equations of state. Appl. Energy 80, 383–399. https://doi.org/10.1016/j.apenergy.2004.04.011

Zehioua, R., Coquelet, C., Meniai, A.H., Richon, D., 2010. Isothermal vapor-liquid equilibrium data of 1,1,1,2-tetrafluoroethane (R134a) + dimethylformamide (DMF) working fluids for an absorption heat transformer. J. Chem. Eng. Data 55, 985–988. https://doi.org/10.1021/je900440t

3 Absorber Configurations

The complexity of vapour absorption refrigeration system (VARS) increases due to various components such as the generator, rectifier, condenser, throttle valve, evaporator, absorber, pump, and solution heat exchanger (SHX). Among these components, the absorber is considered one of the critical elements because of the following reasons:

1. Absorption efficiency directly impacts the VARS performance and cost (Stolk and Waszenaar, 1986; Fatouh and Murthy, 1996; Ibarra-Bahena and Romero, 2014a).
2. Absorber is a liquid-gas contacting system that involves coupled heat and mass transfer between two phases, which is a complex phenomenon that results in more irreversibility due to mixing losses, circulating losses, and imperfect heat and mass transfer.
3. Absorber results in poor solution heat and mass transfer coefficients, resulting in a large-size absorber (Xie et al., 2008).
4. Exergy loss during the second law analysis of VARS is highest for the absorber (Talbi and Agnew, 2000; Şencan et al., 2005; Kilic and Kaynakli, 2007; Anand et al., 2018). This is due to the heat of mixing of the solution, which is absent in pure fluids.

Therefore, this chapter reviews different absorber configurations to overcome the above difficulties.

The function of the absorber is to increase the concentration of the weak solution (absorbent + less percentage of refrigerant) by absorbing the refrigerant vapour (absorbate) coming from the evaporator. Due to the non-ideal behaviour of the working mixtures used in VARS, they liberate heat at the gas-liquid interface during absorption, which is an exothermic reaction. The absorber has to transfer this heat of mixing to the coolant through the solution.

The following key features should be considered when designing an absorber (Selim and Elsayed, 1999a; Ibarra-Bahena and Romero, 2014):

1. The interfacial area between vapour and liquid should be maximum.
2. The heat of mixing has to be removed to sustain the absorption process further.
3. Less pressure drops on the coolant, vapour, and liquid sides.
4. Better heat and mass transfer characteristics between vapour and liquid phases.
5. Cost-effective and compact in size from an economic point of view.

DOI: 10.1201/9781003485193-3

Open literature shows that the heat and mass transfer characteristics of absorbers are poor. To improve these characteristics, researchers have been experimenting with different absorber configurations. Enhancing the absorption process can achieve lower cooling temperatures and smaller total system size. The absorber configurations include a tray, packed bed, falling film (horizontal and vertical), spray, bubble, and membrane absorbers. The following sections discuss the various absorber configurations used in VARS to enhance heat and mass transfer characteristics. The selection of an absorber configuration for a given absorption process depends on multiple parameters such as solution flow rate, required cooling capacity, solution concentration, and the thermophysical properties of the working fluids.

3.1 TRAY OR PLATE COLUMN ABSORBER

The tray column is of the cylindrical tower and consists of several trays or plates that help in increasing the liquid holdup on each plate, thereby improving the liquid residence time. Generally, cross-flow is preferred to enhance its performance in this absorber type. With the increased liquid holdup and the vapour passing through the perforations of the plate, the absorption rate increases due to good contact between the liquid and vapour. Several types of plates are available in the literature, including bubble caps, sieve plates, valve plates, etc. The schematic diagram of the tray absorber is presented in Fig. 3.1.

In general, to determine tray efficiency, the A.I.Ch.E. (1958) model and Colburn's (1936) (Lee et al., 2002) equations are commonly used. However, these models are based on different assumptions that neglect the effect of entrainment, which affects mass transfer in tray columns. Liquid entrainment has a significant influence on tray efficiency. A one-parametric diffusion model is developed to estimate tray efficiency with entrainment (Jaćimović, 2000). This model accurately predicts mass transfer parameters in the presence of entrainment and estimates the apparent tray efficiency when there is no mass transfer in the settling zone. Whenever there is intensive entrainment of the liquid (i.e., the liquid flow rate across the tray is relatively low compared to the gas flow rate), the apparent efficiency of the plate cannot be determined using Colburn's model but can be obtained using the model proposed by Jaćimović and Genić (2008). They modified the previously proposed model of tray efficiency with entrainment by treating the column as a whole. This model estimates the actual number of trays required based on the inlet and outlet concentrations in one phase and the inlet concentration in the other phase. Additionally, Jaćimović and Genić (2008) proposed another method, the tray-to-tray method, to estimate the number of trays required for intense liquid entrainment.

Nanofluids improve the absorption process in tray columns (Torres Pineda et al., 2012). The authors examined the absorption of CO_2 in methanol-based nanofluids (with suspensions of Al_2O_3 and SiO_2) in an acrylic tray column equipped with 12 sieve-type plates. Experimental findings revealed that the maximum absorption rates are improved by up to 10% with nanofluids compared to pure methanol. It is observed that SiO_2 nanoparticles are more effective than Al_2O_3, and the optimal concentration for CO_2 absorption is 0.05 vol% nanofluid, as shown in Fig. 3.2.

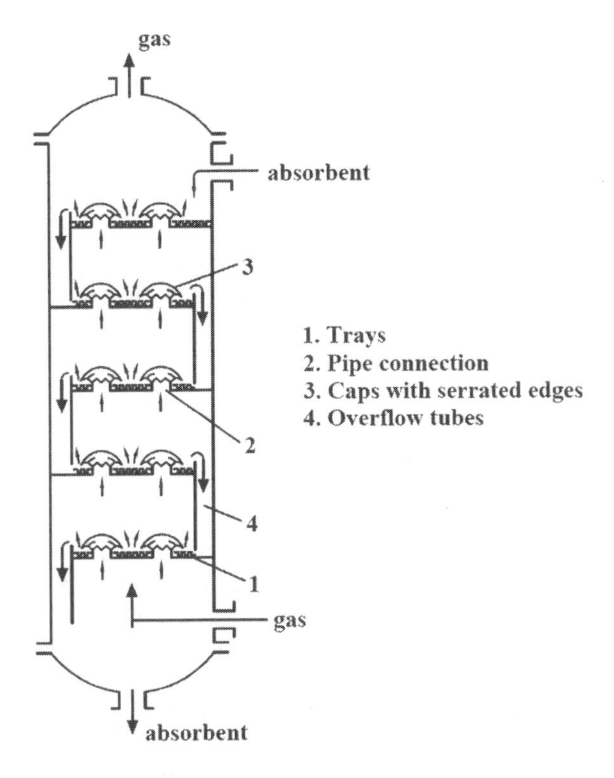

FIGURE 3.1 Tray column (Encyclopedia, 2003).

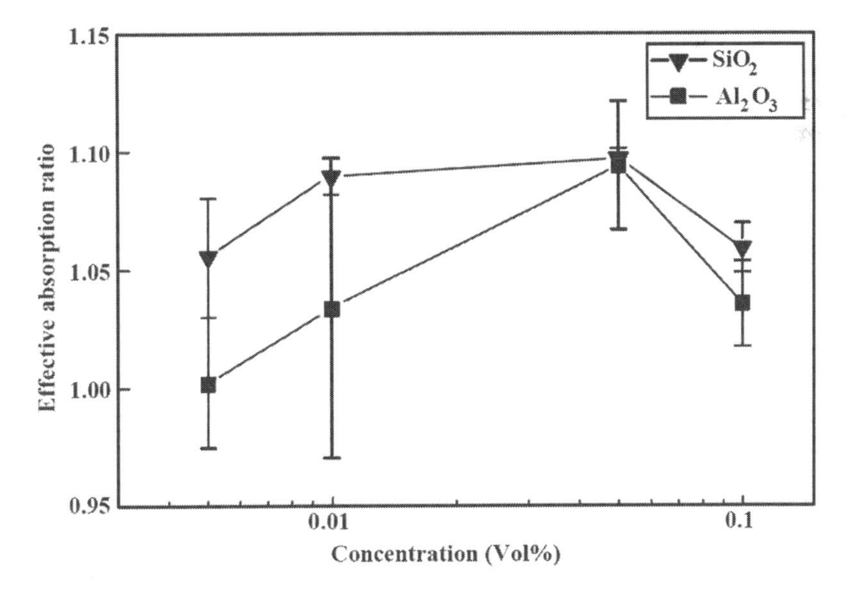

FIGURE 3.2 Effective absorption ratio at different concentrations of SiO_2 and Al_2O_3 (Torres Pineda et al., 2012).

Bubble cap trays are the oldest absorption column technology, consisting of a cap and centre riser. Nevertheless, these have been replaced by sieve and valve trays due to the benefits of these technologies, including ease of operation, high operating range, low maintenance, and lower cost. Sieve plates are circular, rectangular, or square-shaped orifices, while valve plates have movable valves that allow for non-circular cross-section orifices (Perry and Chilton, 1973). Simulation studies on sieve tray columns are carried out to study the absorption of methanol from CO_2 with the help of SIMULINK/MATLAB. This dynamic model considered the effects of perfect mixing of the liquid on plates, negligible vapour holdup, constant total gas flow rate, and linear liquid-vapour equilibrium (Attarakih et al., 2013). Generally, tray types of absorbers are preferred when the absorber diameter is greater than 2 ft, while packed columns are most suitable if the diameter is less than 2 ft due to their lower cost (Perry and Chilton, 1973).

3.2 PACKED BED COLUMN ABSORBER

Packed bed column absorber is a tower-like structure filled with packing materials such as metal, glass, ceramic, etc. These packings can be arranged in a structured or random manner, and it helps to increase the contact area between liquid and vapour. This results in a higher absorption rate as the vapour is diffused more into the liquid. In this configuration, the liquid and vapour flow in a counter-current direction through the packing materials. However, one disadvantage of this absorber is that the heat generated during mixing cannot be removed from the packing materials due to a lack of internal cooling arrangement. Therefore, the packing materials must be replaced periodically, and in this type of absorber, absorption occurs purely due to weak solution subcooling. This absorber type is used in various industries, including absorption, dehumidification using liquid desiccant, rectification, and petrochemicals (Perry and Chilton, 1973; Goel, 2005). When the liquid solution rate is low, packed columns are not preferred because low liquid rate results in incomplete wetting of the packing material, leading to less contact area between the liquid and vapour, resulting in an inefficient absorption process. The packed bed absorber (PBA) is illustrated in Fig. 3.3.

A numerical model is developed to analyse the PBA for an ammonia-water system using raschig rings as the packing material (Selim and Elsayed, 1999a). This model can predict the performance of the bed at different operating and design conditions. The analysis indicated that several parameters affected the absorption process, such as bed height, inlet conditions, mass flow rates of vapour and solution, type of packing material, and volumetric heat rejection. The authors modelled both uniform and non-uniform volumetric heat rejection and found that the non-uniform volumetric heat rejection model provided more practical results. A parametric study is conducted, and the results showed that the operating pressure has a negligible effect on the performance of the bed. As the bed height increases, the transfer area increases, resulting in higher absorption efficiency. However, beyond 0.75m of bed height, the increase in absorption efficiency is minimal, as shown in Fig. 3.4. Increasing the height beyond this height would result in increased pressure drop and bed cost. Therefore, the effective bed height is determined to be 0.75m, which provides an

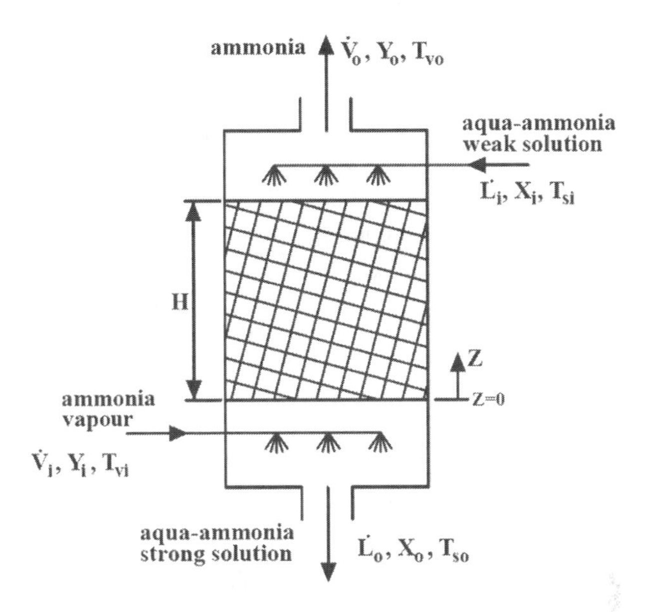

FIGURE 3.3 Packed bed absorber (Selim and Elsayed, 1999a).

absorption efficiency of 94%. Changes in vapour inlet temperature and solution inlet concentration have a negligible effect on absorption efficiency, but there is a slight increase in absorption efficiency due to an increase in solution inlet temperature. Furthermore, the authors replaced the raschig rings with berl saddles of the same equivalent diameter of 25mm and concluded that the berl saddles required less bed height to achieve the same absorption efficiency as the raschig rings.

A numerical simulation is performed to determine the mass transfer coefficient between a mixture of ammonia and water vapour and aqua ammonia solution in a

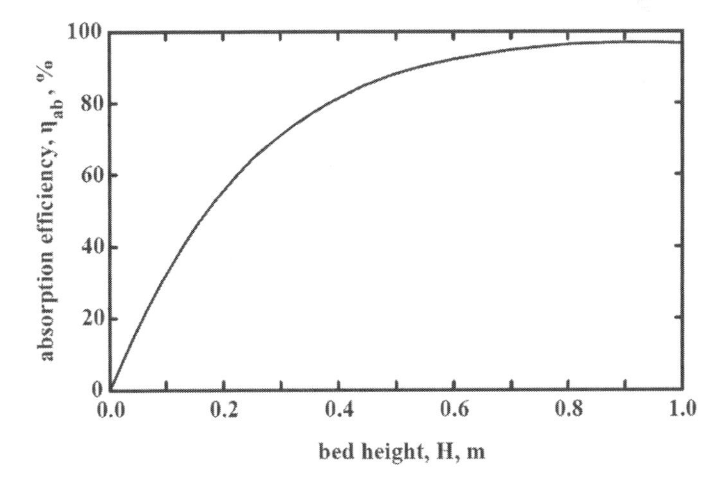

FIGURE 3.4 Effect of bed height on absorption efficiency (Selim and Elsayed, 1999a).

PBA (Selim and Elsayed, 1999b). To account for the mass transfer, the authors have corrected the heat transfer coefficient, and from the simulations, the mass transfer coefficient is derived, which is not directly dependent on interface concentration. The conclusions drawn from the simulation are that the mass transfer flux decreases with an increase in bulk solution concentration, a decrease in solution mass flow rate, a decrease in vapour mass flow rate, and a decrease in the temperature of the liquid solution. Also, the authors have proposed a two-stage PBA with an intercooler to increase the absorption process.

The mass transfer process PBA is influenced by the wetting characteristics of the packing material and interfacial disturbances caused by the Marangoni effect. In a study by Wu (2014), the spontaneous Marangoni effect on mass transfer is experimentally investigated for cylindrical-shaped closely packed polyvinyl chloride PBA, working with water vapour–lithium bromide and water vapour-tri ethylene glycol (TEG) under atmospheric conditions. The Marangoni effect is induced due to the difference in surface tensions between the water vapour and absorbent solutions. The authors concluded that the surface tensions of LiBr and TEG decrease as the temperature increases. The surface tension of the LiBr solution increases as the concentration increases, while that of the TEG solution decreases as the concentration increases, as shown in Fig. 3.5. The authors also found that as the concentration of LiBr and TEG solution increases, the mass transfer coefficient increases. The experimental results showed that the Marangoni effect enhances the mass transfer phenomenon in PBA. For the TEG solution, the mass transfer rate increases when its concentration exceeds 92 wt%.

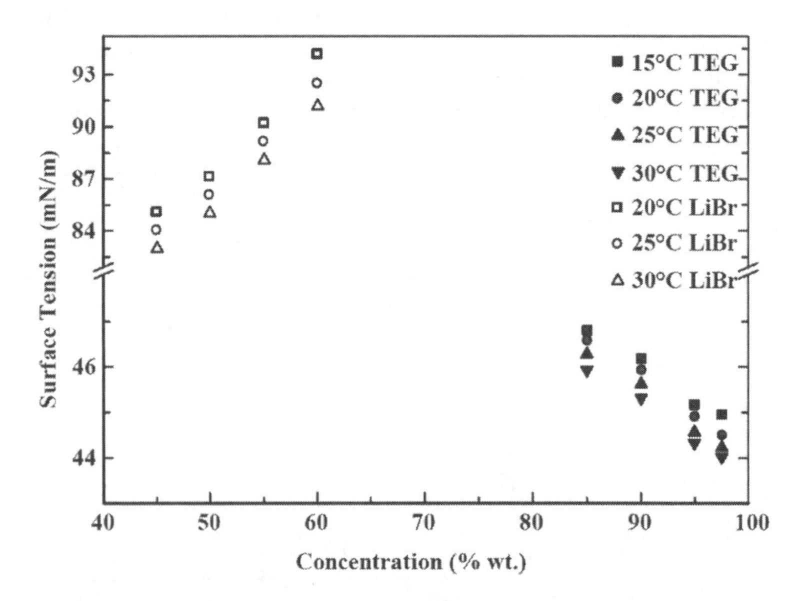

FIGURE 3.5 Effect of concentration and temperature on surface tension for LiBr and TEG (Wu, 2014).

3.3 SPRAY OR ADIABATIC ABSORBER

Unlike the other absorber configurations, in spray absorbers, heat and mass transfer are separated from each other in two different components, that is, the absence of coupled heat and mass transfer. In spray absorbers, heat transfer occurs in an external single-phase heat exchanger, while mass transfer occurs in a simple vessel. This arrangement results in effective heat rejection, better mass transfer, and reduced size and cost. Spray absorbers are also suitable for use with high-viscosity working fluids, such as hydroxides, which have low heat transfer coefficients in falling film absorbers due to their high viscosity and crystallization limits (Summerer et al., 1996; Arzoz et al., 2005). The working fluid is initially subcooled in an external heat exchanger, resulting in a sub-cooled solution. This sub-cooled solution is then sprayed into an adiabatic chamber containing refrigerant vapour by the nozzle. In the adiabatic chamber, the weak solution absorbs the refrigerant vapour until the equilibrium condition is reached, resulting in a strong solution. A schematic diagram of the spray absorber is shown in Fig. 3.6.

An experimental investigation on an adiabatic absorber with a single nozzle is carried out by Warnakulasuriya and Worek (2006). The experiments employed laser technology to quantify the droplet size and velocity. An analytical model based on the Newman model is also developed and found to agree with the experimental results. In experiments, whirl jet nozzles of models 1/8 BX SS 1, 1/8 BX SS 2, 1/8

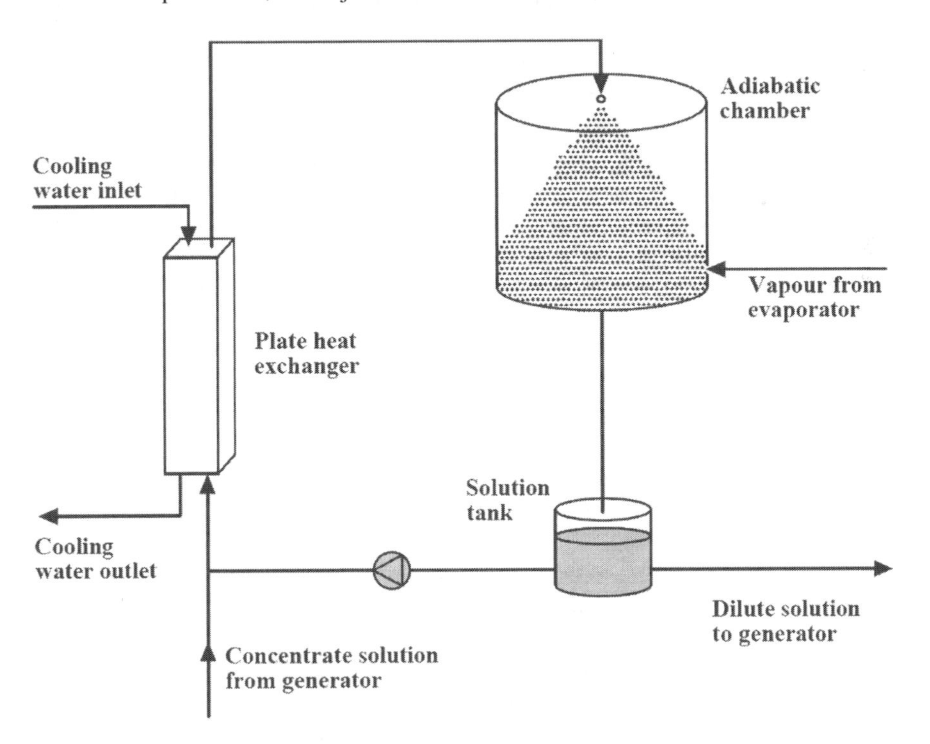

FIGURE 3.6 Spray absorber (Venegas et al., 2004).

BX SS 2W, and 1/8 BX SS 3 and full jet nozzles of models 1/8 GG SS 2 and 1/8 GG SS 2.8W are used. The results indicated that the absorption ratio increases with an increase in flow rate, and the performance of the swirl jet nozzles is superior to that of the other nozzles. Further, Palacios et al. (2009b) conducted experiments on an adiabatic absorber with flat-fan sheets instead of nozzles, which works with LiBr-H_2O.

Experimental evaluation of the adiabatic absorption process of $LiNO_3$-NH_3 using a flat fan nozzle and a sub-cooler is carried out to find the mass transfer rates and performance of the sub-cooler (Zacarías et al., 2011). The results showed that using a flat fan nozzle increased the absorption capacity, and increasing inlet sub-cooling led to an increase in absorption ratio. The absorption ratio is the ratio of vapour absorbed to the mass of the circulated solution. Based on the experimental results, correlations for Sherwood number and approach to equilibrium factor are developed. The same experiments are carried out using a full cone nozzle by Zacarías et al. (2015), a fog jet nozzle by Zacarías et al. (2013), and a fog jet spray by Ventas et al. (2012).

A flat fan sheet adiabatic absorber is modelled and experimented on direct air-cooled single and double-effect LiBr-H_2O absorption machines (González-Gil et al., 2012). The authors concluded that 18 flat fan sheet injectors with a flow rate of 1.6kg s^{-1} could produce 4.5kW cooling in single effect operation mode, while a flow rate of 2.2kg s^{-1} is sufficient to produce 7kW with double effect mode. However, there is a 10% discrepancy between the model and experimental results. In another study, conical liquid sheets are tested for LiBr-H_2O by Palacios et al. (2009a).

The type of spraying arrangement used influences the heat and mass transfer in a spray absorber. The effect of different spraying arrangements, such as mono-dispersed droplets, unstable jets, and falling films using LiBr-H_2O, is studied by carrying out experiments (Arzoz et al., 2005). In all cases, the flow is driven by gravity. The results indicated that the jets and falling film have a faster mass transfer rate due to their higher internal mixing.

Venegas et al. (2005) estimated the mass transfer coefficient by dividing the spray/atomization process into three regions, namely, the liquid jet, drop deceleration, and uniform movement, as shown in Fig. 3.7. In this study, the droplet Reynolds number varied from high Re values at the atomizer outlet to Re 1-10 for uniform movement of the drops. The mass transfer in each spray zone is a function of the interfacial area, contact time between vapour and solution, and mass transfer coefficient. The results indicated that the residence time of droplets in the absorber is determined by the uniform movement zone, which accounted for 86.1% of the total time required to achieve an equilibrium state. In the deceleration period, nearly 60% of the total mass is absorbed, which is 13.4% of the time required to achieve an equilibrium state. The time-averaged mass transfer coefficient achieved is 18.6×10^{-5} ms^{-1}.

Solution drops falling from the nozzle are classified as small droplets (diameter less than 1mm, mostly between 100 and 500μm) and large droplets (greater than 1mm). With small droplets, the absorption rate increases because of the higher surface area, and internal circulation in droplets also helps improve the absorption rate (Warnakulasuriya and Worek, 2006). However, as the diameter of the drop increases, the absorption rate decreases due to decreased surface area. Nevertheless,

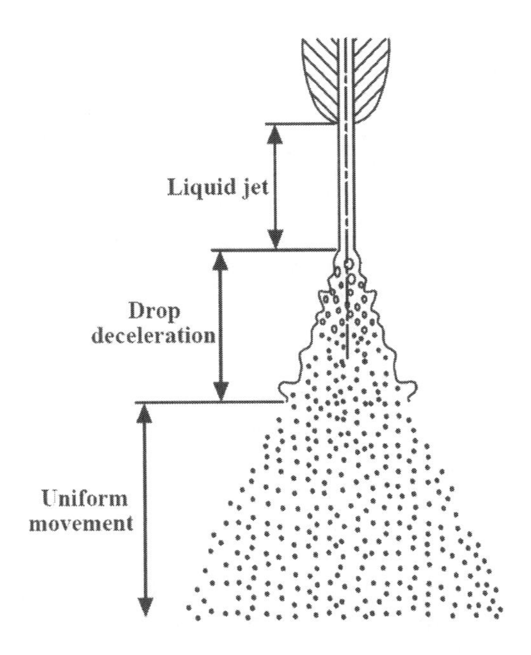

FIGURE 3.7 Different zones in the spray process (Venegas et al., 2005).

the internal circulation increases as the diameter increases, leading to enhanced absorption. This effect must be considered when designing an adiabatic absorber.

The study conducted by Hosseinnia et al. (2016) involved using CFD simulations of large droplets of 4.4mm. The simulations examined the growth, inside velocity, temperature, and concentration profiles of the droplets at different falling times using LiBr–H_2O. As the droplets fall, their shape changes due to surface tension and shear force between the vapour–liquid interface, and they do not reach equilibrium shape and size due to finite absorber size. The absorption rate is higher during the initial stages of the droplets' fall because of a greater mass transfer potential. However, as the droplets fall, the mass transfer potential decreases because of the increased surface temperature of the drop due to absorption heat, hence reducing absorption. The velocity distribution inside the drop is presented in Fig. 3.8, indicating that internal

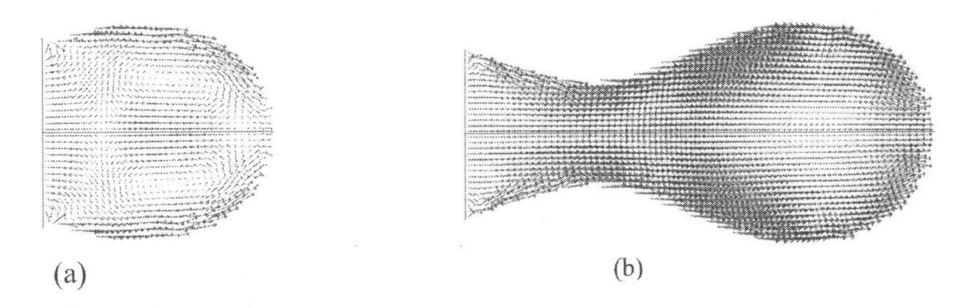

FIGURE 3.8 Velocity vectors inside the drop: (a) sessile drop; (b) pendant drop (Hosseinnia et al., 2016).

circulation is greater in the sessile stage and dampens as the drop grows. In the initial droplet formation stages, mass and heat diffuse from the front stagnation point into the drop, although the mass transfer is lower than heat transfer.

The effect of NH_3–$LiNO_3$ droplet recirculation in an adiabatic and diabatic absorber is numerically carried out by Ventas et al. (2010). The recirculation of the solution helped improve the adiabatic absorber's performance. The efficiency of mass transfer in the absorber chamber mainly depends on droplet diameter, droplet velocity, residence time or length of droplet travel in the absorber chamber, the diffusion coefficient of refrigerant in the solution, amount of sub-cooling, properties of liquid and vapour phase, turbulence intensity, and Reynolds number in a liquid droplet.

Experimental studies on a single-effect LiBr–H_2O adiabatic absorber configuration are carried out with droplets, liquid sheets, and recirculation of diluted solution (Gutiérrez-Urueta et al., 2011). In this study, the absorber's performance is evaluated using two parameters: Efficiency and thermal conductance. The findings indicated that the fan sheets absorber has superior performance characteristics to the droplet absorber, with a size reduction of up to 50%.

The effect of the nozzle size, pressure difference, and solution properties on adiabatic absorption is studied by Warnakulasuriya and Worek (2008). Additionally, the study investigated the effect of drop formation and its size while keeping the spray angle constant at 52° by varying the orifice diameter (SJ1-1.6mm, SJ2-1.98mm, and SJ3-2.39mm). The results showed that when the temperature is kept constant, an increase in nozzle size leads to an increase in drop size for the same pressure difference. Moreover, the study concluded that drop size is directly proportional to the flow rate, which is related to nozzle pressure difference. The study showed that the higher viscosity decreased the spray angle, and an absorption rate improvement of 200% compared to the free-falling film absorber is observed, as depicted in Fig. 3.9.

A mathematical model is developed for the VARS with an adiabatic absorber operating with LiBr–H_2O (Osta-Omar and Micallef, 2016). From a crystallization perspective, the generator temperature is identified as the critical parameter in the design of VARS for LiBr–H_2O. The results of the study showed that the COP increases with an increase in generator temperature or a decrease in adiabatic absorber temperature. For a cooling capacity of 1kW, the recommended generator and adiabatic absorber temperatures are 80°C and 40°C, respectively.

The study by Venegas et al. (2004) involved using numerical simulation to investigate the coupled heat and mass transfer absorption process of a spray absorber with $LiNO_3$–NH_3, specifically in a double-stage VARS with the consideration of solution droplet motion. A detailed literature review of the spray absorber is also included in the article. Smaller droplets increase interfacial area and mass transfer, leading to smaller absorber dimensions and reduced cost. This model determines the concentration profile inside the $LiNO_3$–NH_3 droplets as a function of time. The results indicated that increasing the inlet sub-cooling of the solution droplets led to higher absorption capacity, reduced circulation rates, and lower pump power. The accuracy of the model is compared with the Newman model, and it is found that for droplet sizes of 60 and 100μm, the Newman model is a good predictor of the mass transfer coefficient.

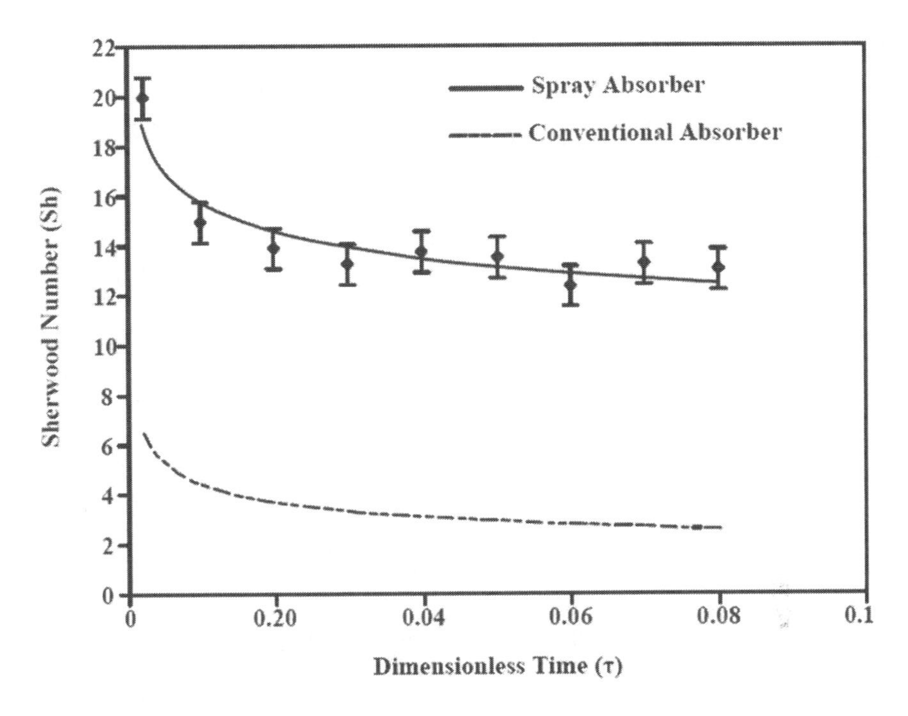

FIGURE 3.9 Comparison of absorption rates of free-falling film and spray absorbers (Warnakulasuriya and Worek, 2008).

The Newman model assumes a constant equilibrium concentration throughout the absorption process, that is the effect of absorption heat is neglected. However, as the process proceeds, the mass transfer potential decreases, causing the equilibrium concentration to change continuously with temperature due to absorption heat, as shown in Fig. 3.10 (Su et al., 2011). To address this, the author developed an improved analytical model of Newman, which considers the effect of absorption heat. According to the findings, this model predicts lower absorption efficiency than the Newman model due to its consideration of absorption heat. For smaller diameter droplets, the absorption rate increases, and absorption heat also increases. Additionally, this model allows for determining the maximum time required for complete absorption, a prediction that the Newman model cannot make.

The experimental testing of a double-effect absorption heat pump with a heating capacity of 45kW is conducted using a mixture of 50% KOH and 50% NaOH as the absorbent with water and hydroxides (Flamensbeck et al., 1998). Although this technique requires more electric power due to increased circulation rates, it can achieve higher heat transfer coefficients in the liquid-liquid heat exchanger. The experiments showed that the mass transfer coefficient in the spray absorber is higher than in the falling film absorber under the same operating conditions. When using a highly efficient pump with an efficiency of 25% or more, the power consumption by the pump is less than 5% of the heating capacity.

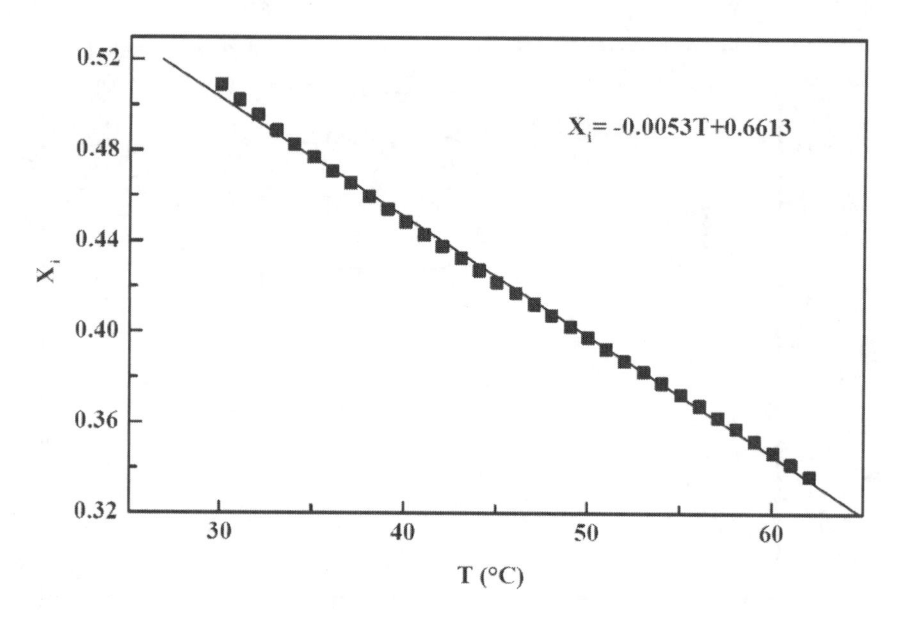

FIGURE 3.10 Equilibrium temperature vs solution concentration at 1333.3 Pa (Su et al., 2011).

An adiabatic ejector is used as an adiabatic absorber in a single effect NH_3–$LiNO_3$ VARS to enhance the performance, increase the refrigerant vapour pressure and reduce costs associated with the solution expansion valve by Vereda et al. (2014). A numerical model is developed to analyse this hybrid cycle, and the results indicated that the adiabatic ejector absorber (AEA) has a lower activation temperature (minimum vapour generation temperature) than both the conventional adiabatic absorber (CAA) and the diabatic absorber (DA), with a difference of approximately 15°C.

The use of an adiabatic absorber in VARS also offers the benefit of miniaturization. Experimental investigations are conducted to attribute this advantage to 5kW and 4.5kW air-cooled single-effect LiBr–H_2O systems with adiabatic absorbers and evaporators for residential applications (Lizarte et al., 2012; Chen et al., 2018). Because of the adiabatic flash evaporation process, both the COP and cooling capacity are increased. Osta-Omar and Micallef (2017) observed the vapour–solution interfacial area in miniaturized LiBr–H_2O systems using a horizontal spiral groove adiabatic absorber made up of perspex material. With all other parameters held constant, the results showed a direct relationship between interfacial area and absorption rate, with an optimum interfacial area of 140 cm^2 for a 45W capacity.

3.4 MEMBRANE ABSORBER

Researchers have recently developed a membrane absorber to control the film thickness. The membrane absorber consists of a constrained film flow, which includes hydrophobic micro-porous membrane contactors located between the refrigerant and weak solution, as shown in Fig. 3.11. The porous membrane is selective and only

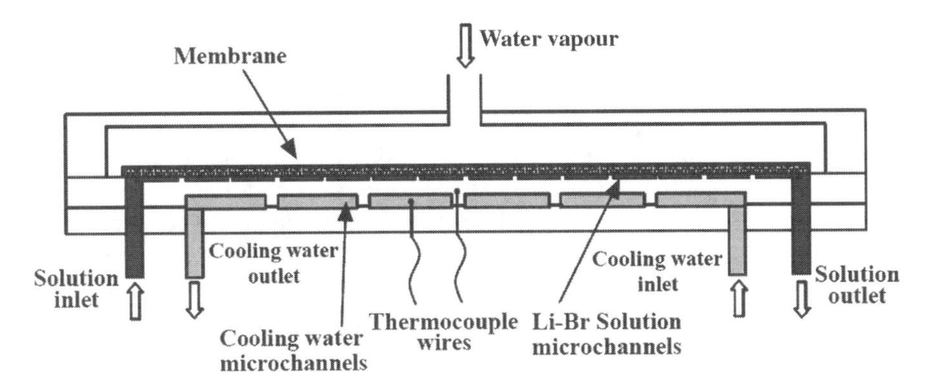

FIGURE 3.11 Membrane absorber (Nasr Isfahani and Moghaddam, 2013).

permits refrigerant vapour to flow through but prevents the solution from flowing. The mass transfer from vapour to solution mainly depends on the membrane properties, such as thickness, porosity, and pore size. This membrane technology is used in absorption processes, separations, distillations, food industries, etc. The use of membrane contactors allows for constructing a compact absorber with a high heat and mass transfer coefficient due to the high interfacial area between vapour and solution. Therefore, it is feasible to miniaturize the entire VARS with this membrane contactor. Membrane technology has the advantage of eliminating fouling problems, but the disadvantages are clogged pores, low mechanical strength, and low operating pressures, limiting its capacity to lower cooling capacity applications (Asfand and Bourouis, 2015).

Numerical and experimental studies are conducted for LiBr–H_2O, and recommendations are made for suitable membrane properties for LiBr–H_2O (Ali and Schwerdt, 2009). The recommended membrane properties include being hydrophobic to LiBr solution with a high entry pressure to prevent pore wettability, no capillary condensation to avoid pore blockage, high permeability to water vapour, a 60 μm thick hydrophobic coating, a porosity of 0.8, and a mean pore size of 0.45 μm. The design of a compact plate and frame absorber with a microporous membrane is carried out by Ali (2010), where it is concluded that counterflow conditions resulted in a more compact absorber. Microporous membrane modelling and parametric studies are carried out for LiBr–H_2O by Venegas et al. (2016b) and Venegas et al. (2016a). These studies are later extended for NH_3–H_2O by Berdasco et al. (2017).

An experiment is conducted by Nasr Isfahani and Moghaddam (2013) to investigate the absorption characteristics of LiBr solution using a superhydrophobic membrane. The effects of solution film thickness, solution flow rate, water vapour pressure, and coolant temperature on absorption are analysed. The results showed that absorption rates could be achieved with high solution flow rates and less solution film thickness, with an improvement of about 2.5 times that of falling film. An absorption rate of 0.006 kg m^{-2} s^{-1} is achieved with a film thickness of 100 μm and a velocity of 5 mm s^{-1}.

CFD simulation studies are carried out on a membrane absorber to study the heat and mass transfer characteristics of LiBr–H_2O (Asfand et al., 2015). The absorption rate is found to be strongly influenced by the thickness and velocity of the solution film. If the solution film thickness is decreased from 2 to 0.5 mm, absorption is improved by a factor of three. However, decreasing the film thickness also led to an increase in pressure drop. Therefore, an optimal solution film thickness has to be determined. The authors concluded that the optimal solution film thickness and velocity are 0.5 mm and 0.005 m s^{-1}, respectively, under the given operating conditions. The effect of membrane properties on absorption rate is numerically studied by Asfand et al. (2016a). It is found that an increase in mean pore size led to an increase in absorption rate but at the cost of selectivity. A 75% improvement in absorption rate is achieved as the mean pore size increased from $0.25\,\text{to}\,1\,\mu\text{m}$.

In a study by Asfand et al. (2016b), CFD simulations are conducted on hydrophobic membranes made of polymers. The simulations are carried out using two different solutions, LiBr + LiI + $LiNO_3$ + LiCl (60.16%:9.55%:18.54%:11.75%)–H_2O and $LiNO_3$ + KNO_3 + $NaNO_3$ (53%:28%:19%)–H_2O to observe the local heat and mass transfer characteristics. The authors found that the absorption rate is 25% higher when using LiBr + LiI + $LiNO_3$ + LiCl compared to using LiBr alone. Additionally, the pressure drop is lower when using $LiNO_3$ + KNO_3 + $NaNO_3$ compared to LiBr and when using LiBr + LiI + $LiNO_3$ + LiCl compared to LiBr.

Numerical studies on a membrane absorber with a rectangular microchannel are carried out under both diabatic and adiabatic conditions by Venegas et al. (2017). The authors analysed the results of various factors, including absorption potential, absorption rate, solution concentration, mass transfer coefficient, and the temperatures of the solution and cooling water. Based on their findings, the authors suggested that adiabatic conditions are most suitable for practical applications because they eliminate the need for a cooling tower and allow for sub-cooling of the solution using atmospheric air.

The following chapters discuss the other absorber configurations, such as falling film and bubble absorbers.

REFERENCES

American Institute of Chemical Engineers. Research Committee (A.I.Ch.E.), 1958. *Bubble-tray Design Manual: Prediction of Fractionation Efficiency.* American Institute of Chemical Engineers.

Ali, A.H.H., 2010. Design of a compact absorber with a hydrophobic membrane contactor at the liquid-vapor interface for lithium bromide-water absorption chillers. Appl. Energy 87, 1112–1121. https://doi.org/10.1016/j.apenergy.2009.05.018

Ali, A.H.H., Schwerdt, P., 2009. Characteristics of the membrane utilised in a compact absorber for lithium bromide-water absorption chillers. Int. J. Refrig. 32, 1886–1896. https://doi.org/10.1016/j.ijrefrig.2009.07.009

Anand, Y., Tyagi, S.K., Anand, S., 2018. Variable capacity absorption cooling system performance for building application. J. Therm. Eng. 4, 2303–2317.

Arzoz, D., Rodriguez, P., Izquierdo, M., 2005. Experimental study on the adiabatic absorption of water vapor into LiBr-H2O solutions. Appl. Therm. Eng. 25, 797–811. https://doi.org/10.1016/j.applthermaleng.2004.08.003

Asfand, F., Bourouis, M., 2015. A review of membrane contactors applied in absorption refrigeration systems. Renew. Sustain. Energy Rev. 45, 173–191. https://doi.org/10.1016/j.rser.2015.01.054

Asfand, F., Stiriba, Y., Bourouis, M., 2015. CFD simulation to investigate heat and mass transfer processes in a membrane-based absorber for water-LiBr absorption cooling systems. Energy 91, 517–530. https://doi.org/10.1016/j.energy.2015.08.018

Asfand, F., Stiriba, Y., Bourouis, M., 2016a. Impact of the solution channel thickness while investigating the effect of membrane characteristics and operating conditions on the performance of water-LiBr membrane-based absorbers. Appl. Therm. Eng. 108, 866–877. https://doi.org/10.1016/j.applthermaleng.2016.07.139

Asfand, F., Stiriba, Y., Bourouis, M., 2016b. Performance evaluation of membrane-based absorbers employing $H_2O/(LiBr + LiI + LiNO_3 + LiCl)$ and $H_2O/(LiNO_3 + KNO_3 + NaNO_3)$ as working pairs in absorption cooling systems. Energy 115, 781–790. https://doi.org/10.1016/j.energy.2016.08.103

Attarakih, M., Abu-Khader, M., Bart, H.J., 2013. Dynamic analysis and control of sieve tray gas absorption column using MATALB and SIMULINK. Appl. Soft Comput. J. 13, 1152–1169. https://doi.org/10.1016/j.asoc.2012.10.011

Berdasco, M., Coronas, A., Vallès, M., 2017. Theoretical and experimental study of the ammonia/water absorption process using a flat sheet membrane module. Appl. Therm. Eng. 124, 477–485. https://doi.org/10.1016/j.applthermaleng.2017.06.027

Chen, J.F., Dai, Y.J., Wang, H.B., Wang, R.Z., 2018. Experimental investigation on a novel air-cooled single effect $LiBr\text{-}H_2O$ absorption chiller with adiabatic flash evaporator and adiabatic absorber for residential application. Sol. Energy 159, 579–587. https://doi.org/10.1016/j.solener.2017.11.029

Encyclopedia, 2003. Absorber. (n.d.) McGraw-Hill Dictionary of Scientific & Technical Terms, 6E (2003). Retrieved April 3 2023 from https://encyclopedia2.thefreedictionary.com/Absorber [WWW Document].

Fatouh, M., Murthy, S.S.A.S.A., 1996. HCFC22-based vapour absorption refrigeration systems. Part ii: Influence of component effectiveness. Int. J. Energy Res. 20, 371–384.

Flamensbeck, M., Summerer, F., Riesch, P., Ziegler, F., Alefeld, G., 1998. A cost effective absorption chiller with plate heat exchangers using water and hydroxides. Appl. Therm. Eng. 18, 413–425. https://doi.org/10.1016/S1359-4311(97)00049-5

Goel, N., 2005. Theoretical and experimental analysis of absorption- condensation in a combined power and cooling cycle. PhD thesis, submitted to University of Florida.

González-Gil, A., Izquierdo, M., Marcos, J.D., Palacios, E., 2012. New flat-fan sheets adiabatic absorber for direct air-cooled LiBr/H2O absorption machines: Simulation, parametric study and experimental results. Appl. Energy 98, 162–173. https://doi.org/10.1016/j.apenergy.2012.03.019

Gutiérrez-Urueta, G., Rodríguez, P., Venegas, M., Ziegler, F., Rodríguez-Hidalgo, M.C., 2011. Experimental performances of a LiBr-water absorption facility equipped with adiabatic absorber. Int. J. Refrig. 34, 1749–1759. https://doi.org/10.1016/j.ijrefrig.2011.07.014

Hosseinnia, S.M., Naghashzadegan, M., Kouhikamali, R., 2016. CFD simulation of adiabatic water vapor absorption in large drops of water-LiBr solution. Appl. Therm. Eng. 102, 17–29. https://doi.org/10.1016/j.applthermaleng.2016.03.144

Ibarra-Bahena, J., Romero, R.J., 2014. Performance of different experimental absorber designs in absorption heat pump cycle technologies: A review. Energies 7, 751–766. https://doi.org/10.3390/en7020751

Jaćimović, B.M., 2000. Entrainment effect on tray efficiency. Chem. Eng. Sci. 55, 3941–3949. https://doi.org/10.1016/S0009-2509(99)00583-7

Jaćimović, B.M., Genić, S.B., 2008. Tray-to-tray method for estimation of the number of trays in gas-liquid columns in case of intensive entrainment. Chem. Eng. Res. Des. 86, 427–434. https://doi.org/10.1016/j.cherd.2007.12.006

Kilic, M., Kaynakli, O., 2007. Second law-based thermodynamic analysis of water-lithium bromide absorption refrigeration system. Energy 32, 1505–1512. https://doi.org/10.1016/j.energy.2006.09.003

Lee, K.B., Chun, B.H., Lee, J.C., Lee, C.H., Kim, S.H., 2002. Experimental analysis of bubble mode in a plate-type absorber. Chem. Eng. Sci. 57, 1923–1929. https://doi.org/10.1016/S0009-2509(02)00089-1

Lizarte, R., Izquierdo, M., Marcos, J.D., Palacios, E., 2012. An innovative solar-driven directly air-cooled LiBr-H2O absorption chiller prototype for residential use. Energy Build. 47, 1–11. https://doi.org/10.1016/j.enbuild.2011.11.011

Nasr Isfahani, R., Moghaddam, S., 2013. Absorption characteristics of lithium bromide (LiBr) solution constrained by superhydrophobic nanofibrous structures. Int. J. Heat Mass Transf. 63, 82–90. https://doi.org/10.1016/j.ijheatmasstransfer.2013.03.053

Osta-Omar, S.M., Micallef, C., 2016. Mathematical model of a lithium-bromide/water absorption refrigeration system equipped with an adiabatic absorber. Computation 4, 1–16. https://doi.org/10.1016/j.egypro.2017.07.009

Osta-Omar, S.M., Micallef, C., 2017. Effect of the vapour-solution interface area on a miniature lithium-bromide/water absorption refrigeration system equipped with an adiabatic absorber. Energy Procedia 118, 243–247. https://doi.org/10.1016/j.egypro.2017.07.009

Palacios, E., Izquierdo, M., Lizarte, R., Marcos, J.D., 2009a. Lithium bromide absorption machines: Pressure drop and mass transfer in solutions conical sheets. Energy Convers. Manag. 50, 1802–1809. https://doi.org/10.1016/j.enconman.2009.03.023

Palacios, E., Izquierdo, M., Marcos, J.D., Lizarte, R., 2009b. Evaluation of mass absorption in LiBr flat-fan sheets. Appl. Energy 86, 2574–2582. https://doi.org/10.1016/j.apenergy.2009.04.033

Perry, R.H., Chilton, C.H., 1973. Perry's chemical engineers' handbook, Fifth ed. Mc Graw-Hill, New York.

Selim, A.M., Elsayed, M.M., 1999a. Performance of a packed bed absorber for aqua ammonia absorption refrigeration system. Int. J. Refrig. 22, 283–292. https://doi.org/10.1016/S0140-7007(98)00066-8

Selim, A.M., Elsayed, M.M., 1999b. Interfacial mass transfer and mass transfer coefficient in aqua ammonia packed bed absorber. Int. J. Refrig. 22, 263–274. https://doi.org/10.1016/S0140-7007(98)00073-5

Şencan, A., Yakut, K.A., Kalogirou, S.A., 2005. Exergy analysis of lithium bromide/water absorption systems. Renew. Energy 30, 645–657. https://doi.org/10.1016/j.renene.2004.07.006

Stolk, A.L., Waszenaar, R.H., 1986. Heat and Mass Transfer Phenomena in an Absorber with Drop-Wise Falling Film on Horizontal Tubes, in: International Refrigeration and Air Conditioning Conference School. pp. 250–256.

Su, F., Ma, H. Bin, Gao, H., 2011. Characteristic analysis of adiabatic spray absorption process in aqueous lithium bromide solution. Int. Commun. Heat Mass Transf. 38, 425–428. https://doi.org/10.1016/j.icheatmasstransfer.2010.12.038

Summerer, F., Riesch, P., Ziegler, F., Alefeld, G., 1996. Hydroxide absorption heat pumps with spray absorber. ASHRAE Trans. 102, 1010–1016.

Talbi, M.M., Agnew, B., 2000. Exergy analysis: An absorption refrigerator using lithium bromide and water as the working fluids. Appl. Therm. Eng. 20, 619–630. https://doi.org/10.1016/S1359-4311(99)00052-6

Torres Pineda, I., Lee, J.W., Jung, I., Kang, Y.T., 2012. CO2 absorption enhancement by methanol-based Al2O3 and SiO2 nanofluids in a tray column absorber. Int. J. Refrig. 35, 1402–1409. https://doi.org/10.1016/j.ijrefrig.2012.03.017

Venegas, M., de Vega, M., García-Hernando, N., Ruiz-Rivas, U., 2016b. A simple model to predict the performance of a H2O-LiBr absorber operating with a microporous membrane. Energy 96, 383–393. https://doi.org/10.1016/j.energy.2015.12.059

Venegas, M., de Vega, M., García-Hernando, N., Ruiz-Rivas, U., 2017. Adiabatic vs non-adiabatic membrane-based rectangular micro-absorbers for H2O-LiBr absorption chillers. Energy 134, 757–766. https://doi.org/10.1016/j.energy.2017.06.068

Venegas, M., De Vega, M., García-Hernando, N., 2016a. Parametric study of operating and design variables on the performance of a membrane-based absorber. Appl. Therm. Eng. 98, 409–419. https://doi.org/10.1016/j.applthermaleng.2015.12.074

Venegas, M., Izquierdo, M., Rodríguez, P., Lecuona, A., 2004. Heat and mass transfer during absorption of ammonia vapour by LiNO3-NH3solution droplets. Int. J. Heat Mass Transf. 47, 2653–2667. https://doi.org/10.1016/j.ijheatmasstransfer.2003.12.014

Venegas, M., Rodríguez, P., Lecuona, A., Izquierdo, M., 2005. Spray absorbers in absorption systems using lithium nitrate-ammonia solution. Int. J. Refrig. 28, 554–564. https://doi.org/10.1016/j.ijrefrig.2004.10.005

Ventas, R., Lecuona, A., Legrand, M., Rodríguez-Hidalgo, M.C., 2010. On the recirculation of ammonia-lithium nitrate in adiabatic absorbers for chillers. Appl. Therm. Eng. 30, 2770–2777. https://doi.org/10.1016/j.applthermaleng.2010.08.001

Ventas, R., Vereda, C., Lecuona, A., Venegas, M., Rodríguez-Hidalgo, M.D.C., 2012. Effect of the NH_3-$LiNO_3$ concentration and pressure in a fog-jet spray adiabatic absorber. Appl. Therm. Eng. 37, 430–437. https://doi.org/10.1016/j.applthermaleng.2011.11.067

Vereda, C., Ventas, R., Lecuona, A., López, R., 2014. Single-effect absorption refrigeration cycle boosted with an ejector-adiabatic absorber using a single solution pump. Int. J. Refrig. 38, 22–29. https://doi.org/10.1016/j.ijrefrig.2013.10.010

Warnakulasuriya, F.S.K., Worek, W.M., 2006. Adiabatic water absorption properties of an aqueous absorbent at very low pressures in a spray absorber. Int. J. Heat Mass Transf. 49, 1592–1602. https://doi.org/10.1016/j.ijheatmasstransfer.2005.11.003

Warnakulasuriya, F.S.K., Worek, W.M., 2008. Drop formation of swirl-jet nozzles with high viscous solution in vacuum-new absorbent in spray absorption refrigeration. Int. J. Heat Mass Transf. 51, 3362–3368. https://doi.org/10.1016/j.ijheatmasstransfer.2007.11.015

Wu, H., 2014. Effect of interfacial phenomena on mass transfer performance of an absorber packed closely with cylindrical packing. Chem. Eng. J. 240, 74–81. https://doi.org/10.1016/j.cej.2013.11.068

Xie, G., Sheng, G., Bansal, P.K., Li, G., 2008. Absorber performance of a water/lithium-bromide absorption chiller. Appl. Therm. Eng. 28, 1557–1562. https://doi.org/10.1016/j.applthermaleng.2007.09.014

Zacarías, A., Venegas, M., Lecuona, A., Ventas, R., 2013. Experimental evaluation of ammonia adiabatic absorption into ammonia-lithium nitrate solution using a fog jet nozzle. Appl. Therm. Eng. 50, 781–790. https://doi.org/10.1016/j.applthermaleng.2012.07.006

Zacarías, A., Venegas, M., Lecuona, A., Ventas, R., Carvajal, I., 2015. Experimental assessment of vapour adiabatic absorption into solution droplets using a full cone nozzle. Exp. Therm. Fluid Sci. 68, 228–238. https://doi.org/10.1016/j.expthermflusci.2015.05.001

Zacarías, A., Venegas, M., Ventas, R., Lecuona, A., 2011. Experimental assessment of ammonia adiabatic absorption into ammonia-lithium nitrate solution using a flat fan nozzle. Appl. Therm. Eng. 31, 3569–3579. https://doi.org/10.1016/j.applthermaleng.2011.07.019

4 Falling Film Absorber

Falling film or wetted column absorbers are widely used in commercial cooling applications. A falling film absorber is a modified version of a packed column, where the packing materials are replaced by heat transfer surfaces such as horizontal or vertical tubes and plates. The weak liquid solution flows downward as a film over tubes by gravity, while the refrigerant vapour flows in the opposite direction to the weak solution and gets absorbed by the solution film. The liberated heat is rejected by the coolant flowing through the tubes. However, the falling film absorber suffers from wettability issues, requiring a liquid distributor to distribute the weak solution evenly over the tubes. Flooding adjacent tubes is also a significant concern with falling film absorbers. Despite these difficulties, falling film absorbers are widely used due to their high heat and mass transfer coefficients and low-pressure drops in vapour and liquid phases (Goel and Goswami, 2005a).

This chapter is divided into two subsections, namely, horizontal and vertical falling film absorbers. The review articles related to falling film absorbers include coupled heat and mass transfer models in falling film absorbers (vertical, horizontal, and inclined) by Killion and Garimella (2001), experimental results of horizontal LiBr–H_2O falling film by Killion and Garimella (2003), and heat and mass transfer correlations for NH_3–H_2O and LiBr–H_2O by Narváez-Romo et al. (2017).

4.1 HORIZONTAL TUBE FALLING FILM ABSORBER

In the horizontal falling film, when the solution is flowing on tubes, it exhibits different flow modes, including droplet, droplet-column, column, column-sheet, and sheet modes, as shown in Fig. 4.1 (Fujita, 1993; Hu and Jacobi, 1996). The falling film flow rate (Reynolds number) significantly influences the transition between these modes. At low flow rates, the liquid leaves the tube in droplet mode, while increasing flow rate leads to column (jet) mode, which merges into the sheet mode at even higher flow rates. Experimental investigations are conducted to identify these transitions, and a transitional Reynolds number in terms of Galileo number is developed. However, the authors have neglected the effects of geometric and thermo-physical properties during the derivation of the transition Reynolds number. Later studies have considered the impact of tube diameter, tube spacing, flow rate, and fluid properties to derive a transition Reynolds number. Finally, the Reynolds number is derived in terms of Galileo number, capillary length, and tube spacing (Wang and Jacobi, 2012).

A 2-D numerical model is developed for LiBr–H_2O to describe the coupled heat and mass transfer in a smooth bundle of horizontal tubes. The authors have used algorithm 688 PDECOL, developed by Keast and Muir (1991), to solve the non-linear partial differential equations within the domain, as shown in Fig. 4.2.

DOI: 10.1201/9781003485193-4

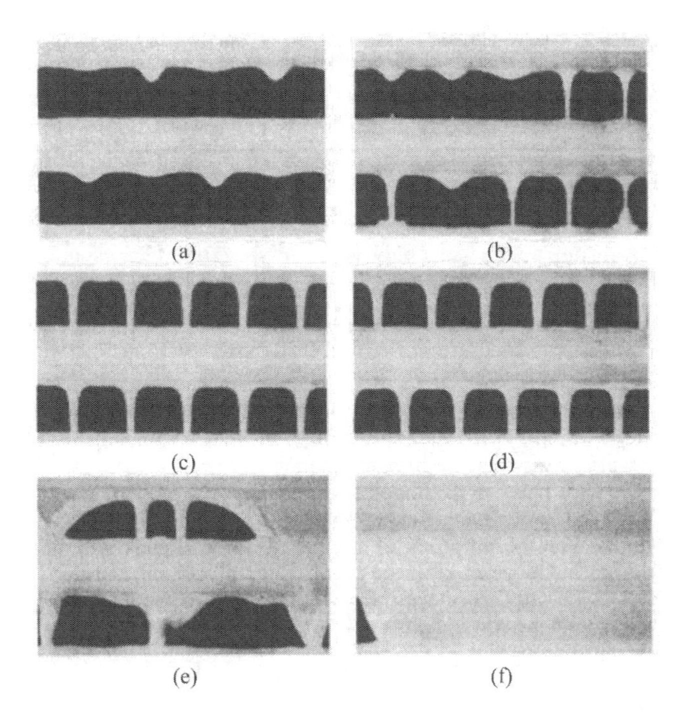

FIGURE 4.1 Types of falling film modes. (a) Droplet; (b) droplet-jet; (c) inline jet; (d) staggered jet; (e) jet-sheet; (f) sheet mode (Hu and Jacobi, 1996).

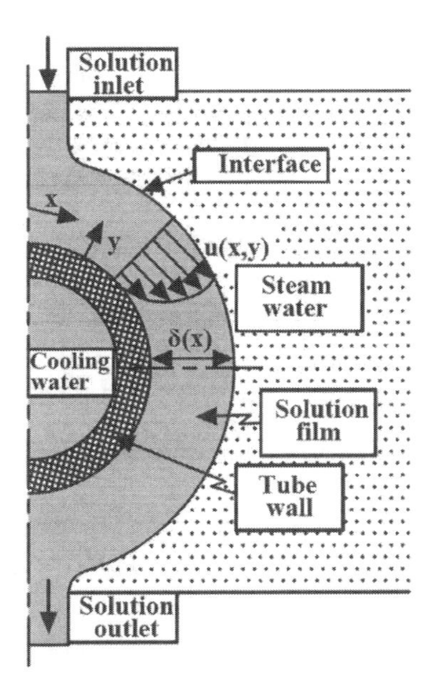

FIGURE 4.2 Cross-sectional view of falling film over horizontal tube (Papaefthimiou et al., 2012).

The results showed that increasing solution mass flow rate leads to a decrease in final temperature of the solution and concentration at exit of the tube bank. However, for solution flow rates above 0.03 kg m^{-1} s^{-1}, the temperature and concentration remain constant (Papaefthimiou et al., 2012). Additionally, the authors found that the maximum efficiency of the absorption process is dependent on the number of horizontal tubes for a given solution inlet conditions and cooling water.

Experimental and numerical analysis of 24 horizontal tubes is carried out to analyse the simultaneous heat and mass transfer process working with LiBr–H_2O (Islam, 2008). The governing equations are solved using the Laplace transformation technique. Results from the parametric revealed that the vapour mass flux increases with an increase in absorber pressure, coolant flow rate, and solution flow rate. Conversely, a decrease in the inlet temperature of the coolant and solution results in increased vapour mass flux, as shown in Fig. 4.3.

A 2-D numerical model is developed to study the heat and mass transfer characteristics of a single horizontal tube working with R134a–DMAC (Harikrishnan et al., 2011b). The model simultaneously solved the energy and mass diffusion equations, assuming laminar flow and 100% wettability. The results showed that concentration increases within the first 40% depth of the film thickness. Furthermore, an increase in solution flow rate results in a decrease in the peak value of mass flux, and its location advances in the flow direction, as shown in Fig. 4.4.

Numerical analysis is conducted on laminar falling film flow in a smooth horizontal tube, taking into account the instantaneous hydrodynamic characteristics of the film (Zhao et al., 2018). The governing equations are discretized using the finite volume method, and a new film thickness correlation is proposed for three regions: 2–15°, 15–165°, and 165–178°. The study revealed that surface tension effects are necessary at the top and bottom stagnation points, and the film thickness increases with a decrease in tube diameter, inlet liquid temperature, and liquid distributor height, as well as with an increase in film flow rate. The authors also found that the

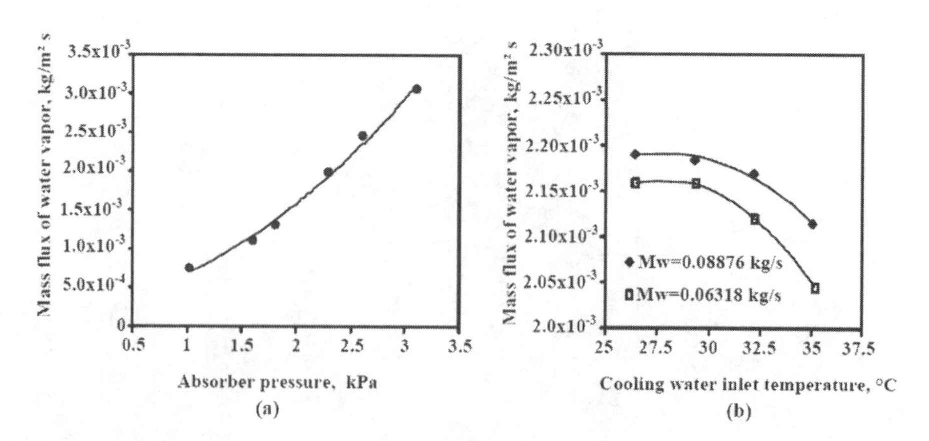

FIGURE 4.3 (a) Effect of pressure on mass flux; (b) effect of coolant temperature and flow rate on mass flux (Islam, 2008).

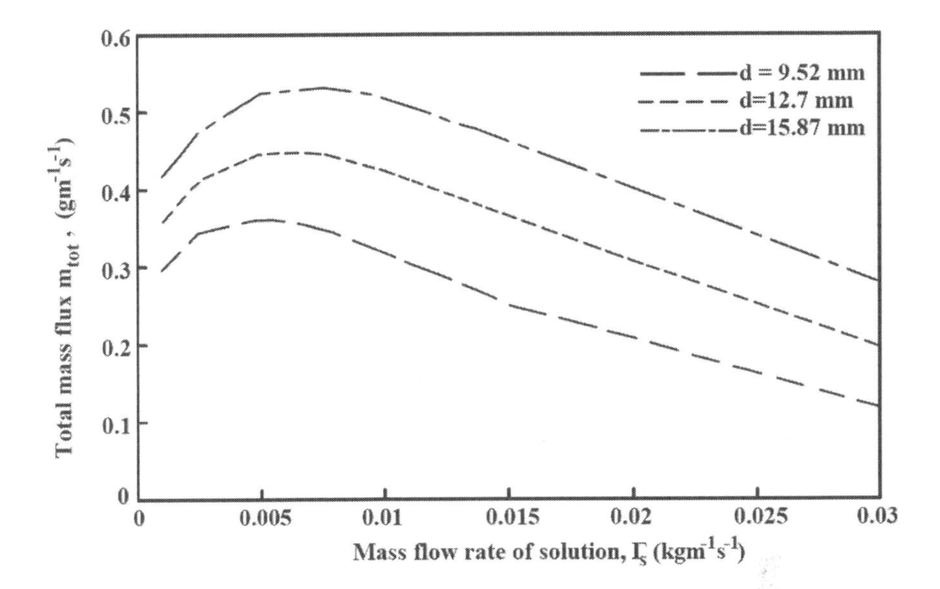

FIGURE 4.4 Variation of mass flux with solution flow rate (Harikrishnan et al., 2011b).

minimum film thickness occurred at a periphery angle of 110–150°, depending on the operating conditions.

The presence of droplets in a horizontal falling film absorber significantly impacts absorption. A 2-D simulation based on an axisymmetric column of spheres is developed to model droplet behaviours, but it failed to predict the effects of interacting droplets and saddle waves (Killion and Garimella, 2004). Therefore, a 3-D simulation is successfully implemented to capture these effects, which are consistent with previous experimental results. The simulation focused on LiBr–H_2O falling film around 15.9 mm diameter tubes with a tube spacing of 15.9 mm. The volume of fluid method is used to define an interface condition with periodic boundary conditions. Also, a 3-D CFD simulation is carried out to observe the drop and jet modes in LiBr–H_2O laminar falling film (Hosseinnia et al., 2017a). The results showed that the drop regime (6.3×10^{-3} $kg\,s^{-1}\,m^{-2}$) has a higher absorption flux than the inline jet mode (4.76×10^{-4} $kg\,s^{-1}\,m^{-2}$). Later, simulations are extended to capture the mixing of solution and the axial propagation of waves caused by droplets impacted by Subramaniam and Garimella (2014). A 3-D simulation on a horizontal tube working with water–air is carried out to visualize the development and elongation of the film along the axial direction (Qiu et al., 2017). This simulation considers the effect of collision and overlap of two adjacent columns of waves in the Reynolds number range of 171–368. The results showed that overlapping two adjacent columns formed like crests, as shown in Fig. 4.5, which significantly influences the film stability and distribution.

A numerical model is developed by Jeong and Garimella (2002) to analyse the behaviour of falling film, droplet formation and droplet free-fall regimes over 13 horizontal tubes for LiBr–H_2O. The model also considered the effect of incomplete

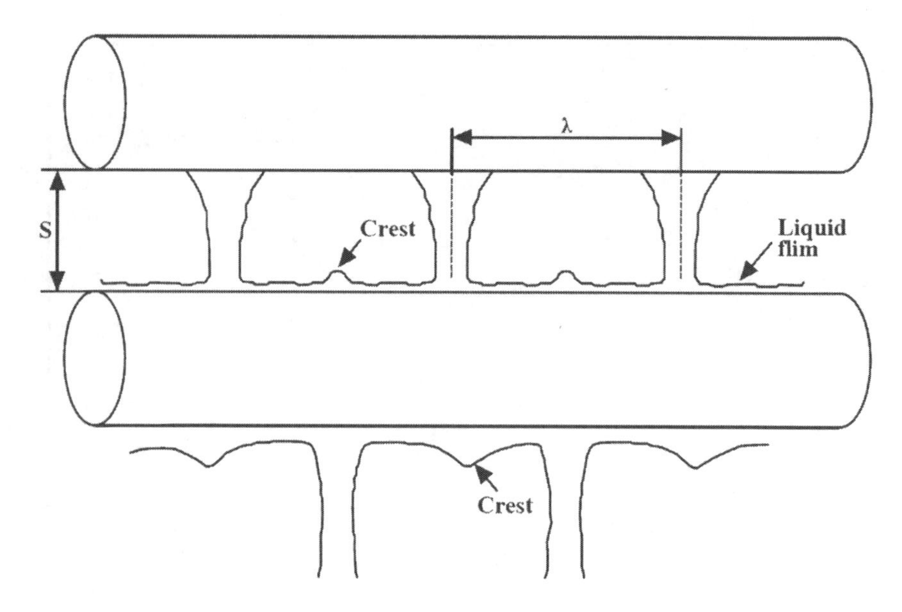

FIGURE 4.5 Columns of the falling film (Qiu et al., 2017).

wetting through the wetting ratio parameter. The study revealed that almost all vapour is absorbed in the falling film and droplet formation regimes. At low solution flow rates, more vapour is absorbed at the top of the tube bank than at the bottom, and more vapour is absorbed in the falling film region than in the droplet formation region. However, as the solution flow rate increases, the droplet region significantly affects heat and mass transfer. The absorption performance reduces when the solution flow rate is below 0.02 kg m^{-1} s^{-1}. Similar studies have also been conducted by Kyung et al. (2007).

In real-time applications, a decrease in mass flow rates may lead to partial wetting problems in falling film and negatively influence the absorption rate. This phenomenon is investigated by a 2-D analytical solution developed for plain horizontal tubes that saves time by avoiding simulations and design (Giannetti et al., 2017). This model is considered partial wetting, cylindrical geometry, and cooling medium temperatures to derive solutions for absorbed mass flux, concentration distribution in film, and Sherwood number. The governing equations are solved by employing the Fourier method, and eigenvalues are derived from the characteristic equation, which depends on Biot, Lewis number, and dimensionless heat of absorption.

Islam et al. (2003) effectively utilized the gap between the tubes to enhance the absorption rate (LiBr–H_2O) by incorporating film inverters (guiding fins). The conventional absorber is modified into an inverting film absorber, as shown in Fig. 4.6. Due to the periodic reversal of falling film direction, this configuration maintains non-equilibrium conditions at the exposed interface, resulting in an enhanced absorption rate. Additionally, the guiding fins prevent the film from forming bands as it flows down the absorber, which was previously observed with the bare tube. Numerical and

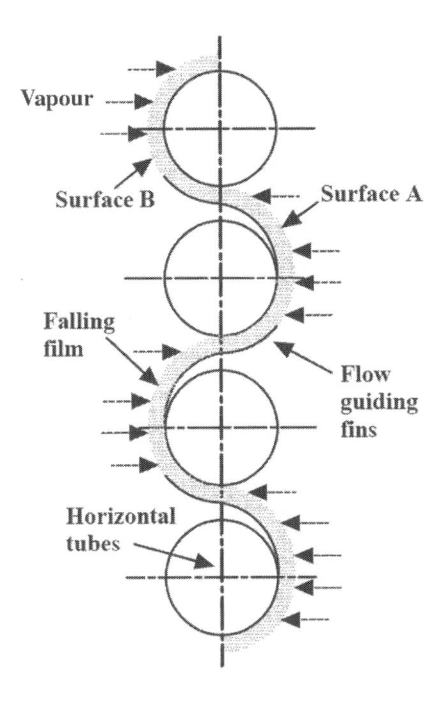

FIGURE 4.6 Film inverted absorber (Islam et al., 2003).

experimental analyses of the falling film absorber demonstrated a maximum increase in absorption rate of approximately 100%. Moreover, the absorption rate increases as film inverting segments increase. In numerical analysis, all 24 horizontal copper tubes are converted into a vertical absorber of height $H = 0.5\pi D_o N + (N-1)\Delta l$ and analysed by laminar falling film model. Subsequently, in 2009, Cui et al. (2009) extended this model for plate falling film and concluded that no additional pressure drop occurred due to the absence of extra guiding vanes in plate configuration.

A novel design for a horizontal falling film is developed by Goel and Goswami (2005a), which utilized the unused vertical space between tubes (working with NH_3–H_2O). In this design, a mesh is wrapped alternatively between the left and right sides of adjacent tubes, as shown in Fig. 4.7. This allowed for the existence of a continuous film between the tubes, which increases the interfacial area and good film stability, resulting in enhanced heat and mass transfer. A numerical 2-D model is developed to analyse simultaneous heat and mass transfer in this absorber configuration, considering mass transfer resistances in both phases. The results showed that the proposed configuration could reduce the size by 25% compared to the traditional horizontal tube design.

In the falling film, velocity, thermal, and concentration gradients exist, leading to a non-equilibrium state, resulting in entropy generation. This phenomenon is studied by an irreversibility analysis conducted for a LiBr–H_2O falling film on a single horizontal tube under steady state, laminar flow, and absence of interfacial waves

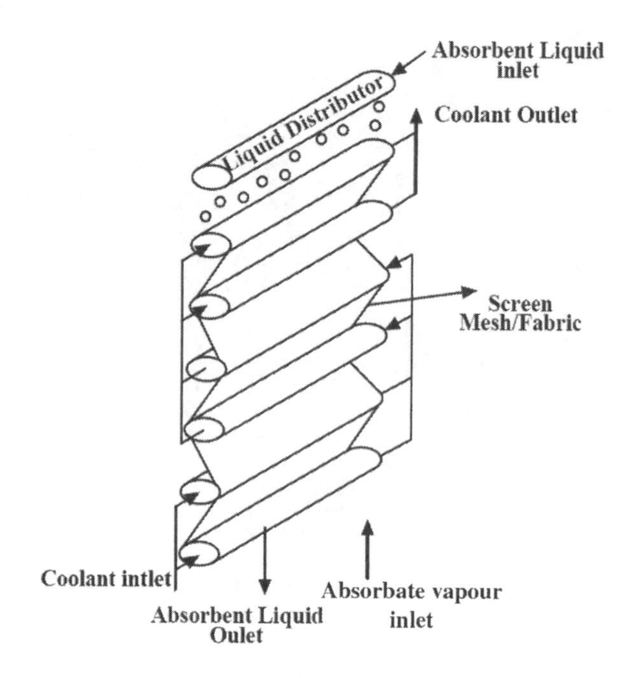

FIGURE 4.7 Falling film absorber with a screen (Goel and Goswami, 2005a).

conditions (Giannetti et al., 2015). Sources of irreversibility are heat transfer (S_t), fluid friction (S_f), coupled heat and mass transfer by convection (S_c), and coupled heat and mass transfer by diffusion (S_d). The study revealed that heat transfer at the wall has a higher impact on entropy generation than that at the liquid–vapour interface. The sources of entropy are discussed, and minimum entropy generation is always in terms of the solution Reynolds number. The overall irreversibility analysis concluded that lower solution flow rates with thin uniform film generate lower entropy. Parametric studies indicated that a lower tube radius, lower coolant flow rate, and temperature enhance the absorption rate and cause high irreversibilities. Therefore, optimization is necessary to find optimal operating conditions that minimize entropy generation. Later, Giannetti et al. (2016) suggested an optimization procedure for the absorber based on volumetric entropy generation.

All of the above studies are related to numerical analysis, and Table 4.1 lists both experimental and enhancement studies.

4.2 VERTICAL SURFACE FALLING FILM ABSORBER

Analytical solutions of temperature and concentration fields are derived from the governing equations by applying film theory and mixture thermodynamics (Kim and Ferreira, 2009). The authors proposed a new approach to calculate the transfer coefficients from these analytical solutions and compared their model to results from previous literature. It is found that conventional methods demonstrated significant inaccuracies in high heat and mass transfer applications. Also, the analytical

TABLE 4.1
Experimental Studies on a Horizontal Falling Film Absorber

References	Working Pair	Absorber Details	Operating Conditions	Results or Major Findings
Killion and Garimella (2003)	Distilled water–air	Horizontal brass tube	D: 12.7 mm No. of tubes: six L: 300 mm Wetted length: 180 mm Tube centre-to-centre distance: 38.1 mm Tube-to-tube distance: 25.4 mm P: 1 atm	The falling film process is captured using a high-speed camera, and detailed regions of the falling film are demonstrated. Droplet, film behaviour, saddle waves resulting from drop impact, non-uniformity, and axial flow characteristics are essential while modelling a falling film absorber.
Yoon et al. (2008)	LiBr–H_2O	Small diameter tubes	OD: 15.88, 12.70, 9.52 mm No. of tubes: 10, 12, 16 P: 0.933 kPa $X_{s,in}$: 61 wt% $T_{s,in}$: 47 °C m_s: 0.0143–0.0303 kg m^{-1} s^{-1} $T_{c,in}$: 32 °C Cooling water velocity: 0.8–1.6 m s^{-1}	The effect of tube diameter on absorber performance is investigated. Among the three tube diameters, the 9.52 mm tube results in a higher heat and mass transfer coefficient of about 9.8% and 11.8% than the 15.88 mm tube. As the diameter of the tube increases, the absorption rate decreases.
Harikrishnan et al. (2011a)	R134a–DMAC	Horizontal tube bundle	X_s: 25.14–42.4% m_s: 0.025–0.045 kg m^{-1} s^{-1} T_c: 20–30 °C m_c: 0.039–0.087 kg s^{-1}	Heat and mass transfer characteristics are studied. An increase in solution flow rate increases absorption rate due to high wetting surface area. An increase in coolant temperature increases the overall heat transfer coefficient while causing a decrease in the mass transfer coefficient.
Soto Francés and Pinazo Ojer (2003)	LiBr–H_2O	Smooth tube	No. of tubes: 14 P: 1.33 kPa X_s: 54–61% T_s: 40–46 °C m_s: 0.01–0.045 kg s^{-1} m^{-2} T_c: 30–37 °C	Experimental results are compared with numerical results.

(*Continued*)

TABLE 4.1 *(Continued)*
Experimental Studies on a Horizontal Falling Film Absorber

References	Working Pair	Absorber Details	Operating Conditions	Results or Major Findings
Deng and Ma (1999)	LiBr–H_2O	Smooth tube	No. of tubes: 24 X_s: 60–63.8% Re_s: 13–39 T_c: 30–32 °C	The effects of cooling water temperature and inlet solution concentration are studied. Reducing coolant temperature from 32 °C to 30 °C results in a 17% increase in heat flux.
Bredow et al. (2008)	LiBr–H_2O	Horizontal copper finned tubes	No. of tubes: 576 OD: 19 mm ID: 16.35 mm L: 1 m m_s: 1800–3000 kg h^{-1} $T_{c,in}$: 16–28 °C	The average heat transfer coefficient is approximately 2000 W m^{-2} K^{-1}, and the effect of the Reynolds number on the heat transfer coefficient is insignificant. The mass transfer coefficient ranges from 50–300 g m^{-2} s^{-1}, depending on solution mass fraction and sub-cooling. It is independent of Re. A weak correlation existed between the Nusselt number and the Prandtl number, as well as between the Sherwood number and Schmidt number.
Álvarez and Bourouis (2018)	Aqueous–nitrate (Lithium, Potassium and sodium) solution	Copper tube bundle with mechanical treatment	Salts ratio: 53:28:19 of $LiNO_3$, KNO_3, and $NaNO_3$ No. of tubes: six OD: 16 mm Thickness: 1 mm L: 400 mm Tube spacing: 30 mm P: 30–35 kPa $X_{s,in}$: 75–82 mass% Re_f: 7–26 $\Delta T_{sub,in}$: 12 °C Re_c: 11,330–17,670 $T_{c,in}$: 70–85 °C	The absorption rate and heat and mass transfer coefficients increase with an increase in cooling water and solution flow rates, absorber pressure, and solution inlet concentration. Conversely, they decrease as the cooling water temperature increases. Under the given operating conditions, the obtained results are as follows: Absorption rate of 2.83–6.55 g m^{-2} s^{-1}, heat transfer coefficient ranging from 631.9 to 1715.8 W m^{-2} °C^{-1}, and mass transfer coefficient between 2.1 and 6×10^{-5} $m s^{-1}$.

TABLE 4.1 *(Continued)*
Experimental Studies on a Horizontal Falling Film Absorber

References	Working Pair	Absorber Details	Operating Conditions	Results or Major Findings
Goel and Goswami (2007)	NH_3–H_2O	Microchannel tube banks with mesh screen	P: 2.81 bar X_s: 0.3 T_s: 43 °C m_s: 10–30 g min^{-1} T_v: 58 °C T_c: 20–30 °C m_c: 269 g min^{-1}	With a mesh screen, the UA value is improved by up to 50% than without a mesh screen. Screen mesh can enhance the mass transfer area and wetting phenomenon.
Park et al. (2003)	LiBr–H_2O	Microscale hatched (Helically grooved) copper tubes	Roughness: 0.386–11.359 μm OD: 14.9–16 mm P: 0.94 kPa $X_{s,in}$: 55–61 wt% $T_{s,in}$: 36–46 °C Re_s: 7–60 $T_{c,in}$: 24–32 °C	Microscale hatched tubes enhanced the absorption process twice more than bare tubes due to increased wettability. Nusselt number correlation is developed in terms of film Reynolds number and surface roughness with an error band of ±25%. Wettability correlation is developed for smooth and roughened tubes (Kim et al., 2003).
Nagavarapu and Garimella (2013)	NH_3–H_2O	Falling film with microchannel tube arrays	X_s: 0.2–0.4 m_s: 0.01–0.024 kg s^{-1} T_v: 40–55 °C T_c: 20–40 °C m_c: 0.139 kg s^{-1}	Heat and mass transfer phenomenon is studied
Kyung and Herold (2002)	LiBr–H_2O	Horizontal tube bank with 2-Ethyl -1-Hexanol	Additive conc.: 0–500 ppm X_s: 0.57–0.6 m_s: 0.01–0.045 kg s^{-1} T_c: 30 °C	1.67 times higher heat transfer coefficient is achieved in the presence of surfactant.
Kang and Kashiwagi (2002)	NH_3–H_2O	Horizontal tube with n-octanol	Additive conc.: 0–800 ppm P: 200 kPa X_s: 0.–0.15 T_s: 18–21 °C	Marangoni convection is visualized. 3–4.6 times improvement of heat transfer is achieved with additive, and an increase in additive concentration results in an improved absorption performance.

(Continued)

TABLE 4.1 *(Continued)*
Experimental Studies on a Horizontal Falling Film Absorber

References	Working Pair	Absorber Details	Operating Conditions	Results or Major Findings
Hoffmann et al. (1996)	LiBr–H_2O	Smooth and knurled tubes with 2E1H and 1-octanol	Additive conc.: 10–640 ppm X_s: 0.4–0.61 T_s: 40–43 °C m_s: 0.3–1.3 l min^{-1}	With knurled tubes and 20–40% additives, the heat transfer coefficient is enhanced by 60–140%.
Park et al. (2004)	LiBr–H_2O	Microscale hatched tube with n-octanol	Roughness: 0.39–697 μm Additive conc.: 400 ppm No. of tubes: 24 P: 0.94 kPa X_s: 0.55–0.61 T_s: 36–46 °C T_c: 24–32 °C	Adding an additive resulted in a 3.76 times higher absorption rate than bare tubes while combining an additive and micro-scale hatched tubes led to a 4.5 times higher absorption rate than bare tubes.
Yoon et al. (2002)	LiBr–H_2O	Horizontal tubes of bare, hydrophilic and floral with surfactants.	Surfactant: n-octanol Surfactant concentration: 500–5500 ppm Tube OD: 15.88 mm Tube ID: 14.05 mm No. of tubes: 48 X_s: 60 wt% $T_{s,in}$: 45 °C m_s: 0.01–0.034 kg m^{-1} s^{-1} $T_{c,in}$: 32 °C Coolant velocity: 1 m s^{-1}	In the absence of surfactants, the hydrophilic tube exhibited a superior heat transfer coefficient compared to the bare tube and floral tubes by 10–35% and 5–25%, respectively. The floral tube demonstrated a better heat transfer coefficient in the presence of surfactants than the bare and hydrophilic tubes. With the addition of surfactants, the heat transfer coefficient of the bare, hydrophilic, and floral tubes is increased by 35–90%, 30–50%, and 40–70%, respectively. Additionally, a surfactant concentration of more than 3500 ppm has an insignificant effect on heat transfer.

TABLE 4.1 *(Continued)*
Experimental Studies on a Horizontal Falling Film Absorber

References	Working Pair	Absorber Details	Operating Conditions	Results or Major Findings
Yang et al. (2011)	NH_3–H_2O	Cylinder with nanoparticles	Nanoparticles: Al_2O_3, Fe_2O_3, and $ZnFe_2O_4$ Nanofluids concentration: 0.1–1.5% Cylinder dimensions: 1200 mm height, 300 mm ID X_s: 0–15%	At a 15% ammonia concentration, Fe_2O_3 and $ZnFe_2O_4$ increase the absorption ratio by 70% and 50%, respectively. The absorption enhancement is mainly due to a decrease in nanofluid viscosity and an improvement in heat transfer. Adding optimal mass fraction of nanoparticles results in a good absorption rate with minimal increase in the nanofluid viscosity.
Kim et al. (2012a)	LiBr–H_2O	With nano-particles	Nano particles: Fe and CNT Nano particle conc.: 0–0.1 wt% P: 0.01 bar X_s: 0.55 T_s: 40 °C T_c: 24–28 °C	An average mass transfer enhancement rate of 2.16 times is achieved with 0.01 wt% of CNT and up to 1.71 times with 0.01 wt% of Fe.
Kim et al. (2012b)	LiBr–H_2O	Copper tubes with nanoparticles and surfactants	Nanoparticles: SiO_2 Conc. of nanofluid: 0–0.01% (vol) Surfactant: 2E1H Conc. of surfactant: 150 ppm L: 500 mm D: 15 mm P: 0.01 bar $X_{s,in}$: 53% $T_{s,in}$: 40 °C m_s: 0.07–0.11 kg s^{-1} $T_{c,in}$: 25 °C	0.01 vol% of SiO_2 concentration is recommended for stability purposes without adding a stabilizer. Compared to combining nanofluids and surfactants, performance enhancement is higher for nanofluids alone. This is because the surfactants' presence inhibited the nanoparticles' Brownian motion. A 46.8% and 18% improvement in heat and mass transfer rates are achieved with 0.005 vol% of SiO_2 nanoparticles.

solutions are used to derive expressions for the effectiveness of heat and mass transfer (Kim and Ferreira, 2010).

Heat and mass transfer occurring at the entrance region of absorbers, desorbers (generators), condensers, and evaporators are mathematically investigated using the self-similarity method (Nakoryakov et al., 2011). The investigation considered an axisymmetric semi-infinite falling film and vapour interface, assuming a uniform film velocity. The findings revealed that the heat and mass transfer process at the entrance region is influenced by various dimensionless numbers such as Froude number (Fr), phase transition criterion for absorption (Ka), Lewis number (Le), and Peclet number (Pe). The study also presented relative variations in temperature and concentration across different cross-sections of the film thickness. Using the self-similarity solution avoids inconsistencies associated with boundary conditions at $x = 0$ and $y = 0$.

Meyer (2014) employed the Laplace transform method to solve the partial differential equations that arise in studying laminar falling films with uniform velocities. The Inverse Laplace transform determines analytical solutions for temperature and concentration. Later, this analysis is extended to account for a realistic boundary condition, that is, a diabatic wall (Meyer, 2015). The resulting temperature and concentration profiles in the film along the length of the vertical surface are presented for two different thermal resistances of the wall and modified Stefan numbers. The absorption flux is greater in the isothermal wall case than in the adiabatic wall case, and doubling the modified Stefan number from 0.1 to 0.2 results in large interfacial concentration gradients. Later, this model is extended to a realistic linear velocity profile by Mortazavi and Moghaddam (2016). The use of a uniform velocity profile leads to an underestimation of the temperature and absorption rate and an overestimation of the concentration profile. However, this model results in approximately 30% error in the absorption rate. All of these models are developed using cartesian coordinates. The same Laplace technique is implemented in cylindrical coordinates by considering the uniform film thickness by Wu (2016).

Based on experimental data, a linearized coupled analytical model is developed to predict heat and mass transfer coefficients in horizontal and vertical tubular absorbers (Islam Md et al., 2004). The authors found that over a shorter distance, the predictions of the linear model are lower than those of the non-linear model, while over longer distances, the linear model predictions are higher. This is attributed to the assumption of the average heat and mass transfer coefficients in the linearized model. However, the linearized model results are comparable to those from the non-linear model with little effort. In some cases, the developed linear model is found to reduce to the logarithmic mean temperature difference (LMTD) method with a 10% error in heat transfer coefficient, but there is considerable discrepancy in the mass transfer coefficient.

LMTD and arithmetic mean temperature difference (AMTD) methods of designing absorption are discussed by Fujita and Hihara (2005). The authors discovered that conventional methods (LMTD and AMTD) divide the absorption process into two independent processes: Heat transfer from solution to coolant and absorption of refrigerant from the vapour phase to solution. However, the falling film mode couples heat and mass transfer, so the conventional methods often produce incorrect results, except for the linear solution temperature profile. Therefore, the authors developed a new method by considering the linear coolant temperature profile, solution concentration, and interfacial equilibrium temperature.

A 1-D numerical model is developed to predict the maximum possible theoretical values of mass transfer enhancement for LiBr–H_2O over a vertical smooth wall surface (Perez-Blanco, 1998). This model solves a system of ordinary differential equations using a fourth-order Runge–Kutta method. The study discovered that the maximum possible mass transfer rate with complete mixing at the interface for a given operating condition is 0.049 kg m^2 s^{-1}.

A numerical model is developed to study the coupled heat and mass transfer in a co-current, in-tube, vertical falling film for NH_3–H_2O (Aminyavari et al., 2017). The governing equations are discretized by the finite difference method. The model is validated with experimental results from the absorption heat transformer. The absorber consists of 37 tubes, each 1 m long, with an outer diameter of 15.8 mm and an inner diameter of 13.8 mm. The study found that heat and mass transfer resistance is more dominant between gas and interface. The majority of the absorption occurred at the top of the tube due to the higher mass transfer potential, as depicted in Fig. 4.8.

The wave characteristics of falling film in a circular tube are numerically investigated for a range of 2–100 mm for tube diameter and 20–100 for Reynolds number, and a theoretical model is developed to describe the free surface deflection (Ho et al., 2001). The governing equations derived are similar to those for a flat plate with

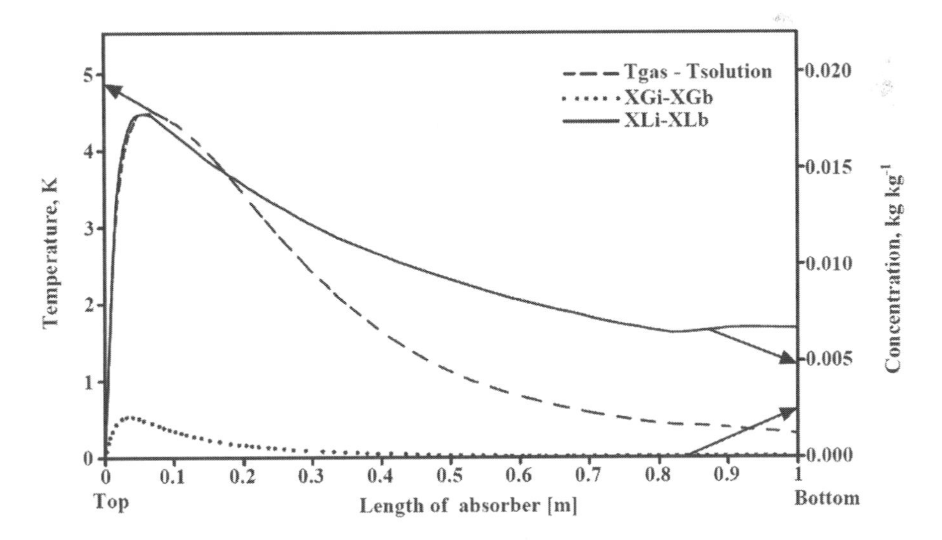

FIGURE 4.8 Driving forces for heat and mass transfer along the length of the absorber (Aminyavari et al., 2017).

an infinite radius. The authors identified wave number, wave velocity, tube radius, and Reynolds numbers are the influencing parameters of wave characteristics and discussed its effects. The study found that as the Reynolds number increases, wave frequency and wave amplitude increase but in a decreasing manner. Furthermore, for the film thickness to tube diameter ratio greater than 4%, a cylindrical model is recommended over the flat plate model.

A numerical model is developed to study the combined heat and mass transfer in a vertical plate absorber for LiBr–H_2O cooled with water (Yoon et al., 2005b). The model made assumptions of laminar flow and fully developed conditions, with no heat transfer in the vapour phase, as depicted in Fig. 4.9. The model enables the calculation of temperature and concentration profiles both along and across the flow direction. The results showed that the highest temperature exists at the interface, while the lowest is near the wall. Conversely, the concentration profile exhibited the opposite trend, with the highest concentration at the wall and the lowest at the interface. The maximum heat and mass transfer coefficients are obtained at a distance of 1.67×10^{-3} m from the inlet, with values of $3.5386 \times 10^{3}\ \mathrm{W\,m^{-2}\,K^{-1}}$ and $1.7527 \times 10^{-4}\ \mathrm{m\,s^{-1}}$, respectively. As the length increases, both transfer coefficients decrease after this length. The study also examined the effect of sub-cooling, superheating, and operating parameters on the absorption performance.

In a subsequent study, Goel and Goswami (2005b) numerically analysed the simultaneous heat and mass transfer in NH_3–H_2O by considering the mass transfer

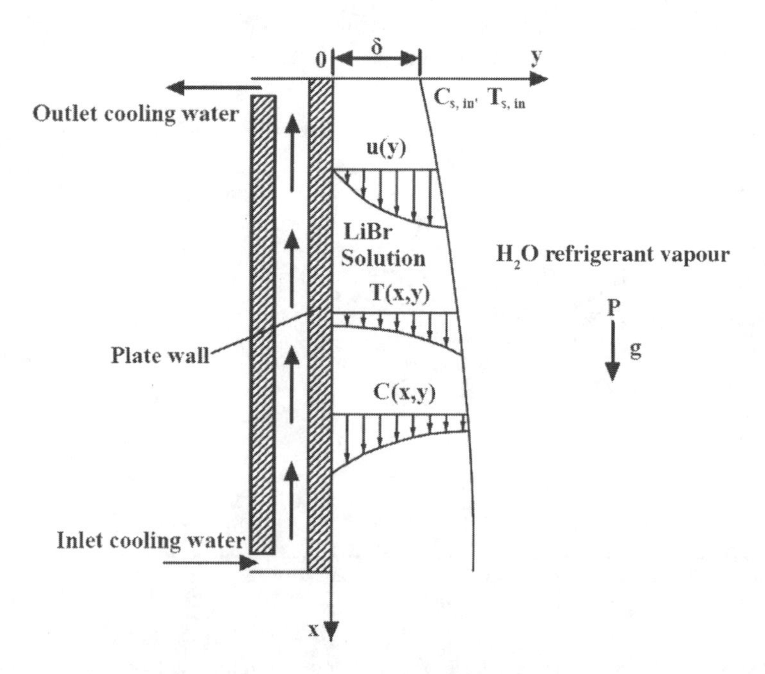

FIGURE 4.9 Schematic model of vertical plate falling film (Yoon et al., 2005b).

resistance in both vapour and liquid phases. The results showed that the molar concentration of ammonia is initially greater than one near the inlet region due to the diffusion of water from the solution to the refrigerant vapour, followed by a gradual decrease below one. The results showed that heat transfer resistance is dominant in the vapour phase, while mass transfer resistance is dominant in the liquid phase. The simultaneous modelling of heat and mass transfer for NH_3–H_2O is numerically investigated using the Colburn analogy, which considers the mass transfer resistance in both phases (Triché et al., 2016, 2017).

Numerical modelling of an inclined plate absorber working with LiBr–H_2O for a range of Reynolds numbers between 5 and 150 is carried out by Karami and Farhanieh (2011). The effect of plate angle on the average Nusselt number, Sherwood number, and Reynolds number is investigated and correlated with the plate angle, as shown in Fig. 4.10. The findings showed that the optimal plate angle for maximum absorption rate is between 85° and 90°. For plate angles greater than 10° and at low Reynolds numbers, the average Nusselt number decreases as the Reynolds number increases due to less dominance of thermal diffusion. However, the Nusselt number increases as the plate angle increases for high Reynolds number flows.

A study on the absorption characteristics of LiBr–H_2O on finned inclined tube surfaces is numerically modelled by Seo and Cho (2004). The authors have investigated the coupled heat and mass transfer characteristics of various tube configurations, including flat, circular, elliptic, and parabolic shapes, as shown in Fig. 4.11. The authors developed a 3-D numerical model to analyse temperature and

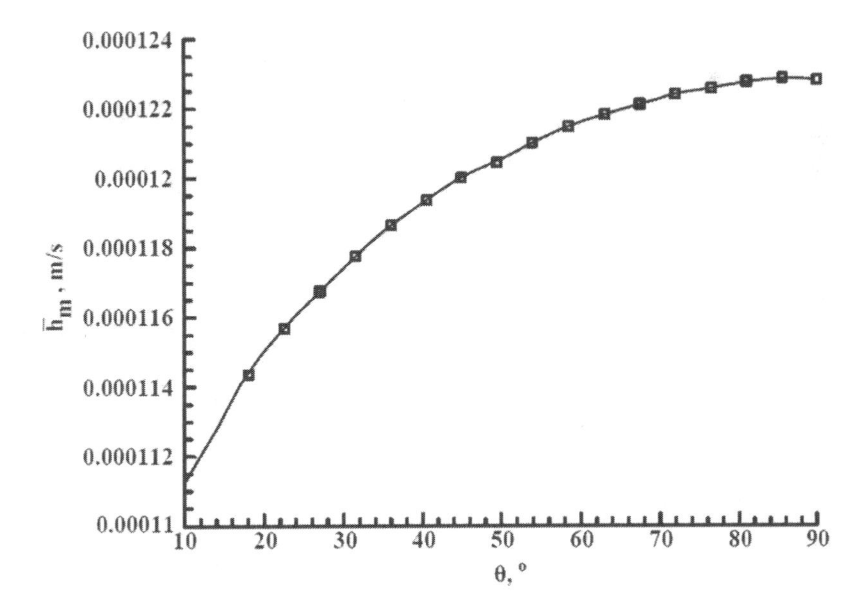

FIGURE 4.10 Effect of plate angle on average mass transfer coefficient at $Re = 20.5$ (Karami and Farhanieh, 2011).

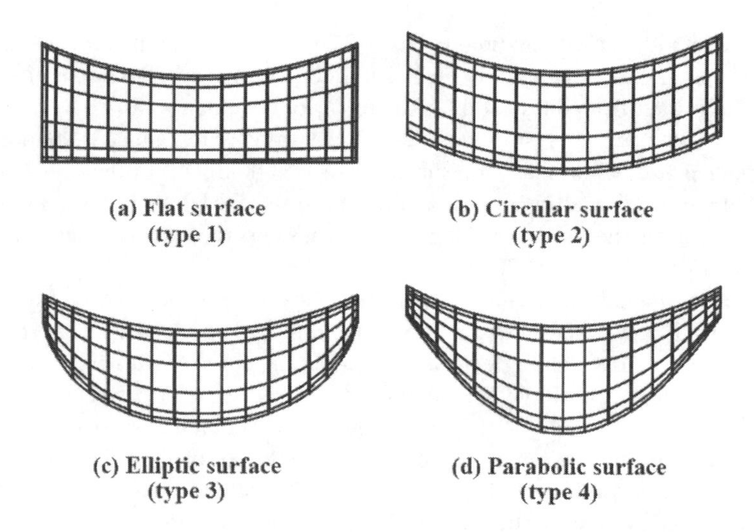

FIGURE 4.11 Types of tube surfaces (Seo and Cho, 2004).

concentration profiles and also examined the effect of fin interval and Reynolds number. The study revealed that maximum absorption occurred at the tip of the fin due to the generation of large temperature and concentration gradients near the fin tip. It is recommended that the film thickness near the fin tip should be thin rather than between the fins to enhance the absorption rate. Based on the results, the authors concluded that the parabolic-shaped tube exhibited better performance characteristics.

The effect of wave motion on low Reynolds number falling film is analysed numerically by employing the Alternating Direction Implicit finite difference method (Yang and Jou, 1995). The results showed that the improvement of temperature and concentration profiles is not a result of the mixing of waves, but instead, it is due to the presence of a vertical component velocity, which is absent in the smooth flow regime.

The effect of waves on heat and mass transfer rates through both experimental and numerical investigations is conducted by Kim and Cho (2005). The numerical approach assumed laminar flow and disregarded mass transfer resistance in the vapour phase. Temperature and concentration profiles are obtained after solving the governing equations. In the experiments, bare, grooved, and spring tubes are tested for operating conditions of Re: 50–150, absorber pressure of 1.01 kPa, and coolant flow rate of 0.025 kg s^{-1}. The maximum wave amplitude of 0.319 mm is achieved for wavelengths greater than 0.02 m, and the wavy flow has a 3% higher maximum wave amplitude than the uniform film flow. The study found that grooved and spring tubes resulted in higher mass and heat transfer rates than bare tubes.

The performance of solitary wave falling film over a vertical surface is numerically investigated for LiBr–H_2O (Islam et al., 2009). The solitary waves are characterized by an asymmetric protrusion with a steep front and gently sloping tail, preceded by front-running capillary ripples. These solitary waves are generated by periodically disturbing the inlet with a specific frequency. The results demonstrated that the solitary waves induced recirculation in the film, as shown in Fig. 4.12, which helps to enhance the mass transfer. The recirculation flow removed the hot and weak solution from the interface and replaced it with a cool and rich LiBr solution to the interface. Also, it showed better heat and mass transfer characteristics than smooth film flow under the same operating conditions.

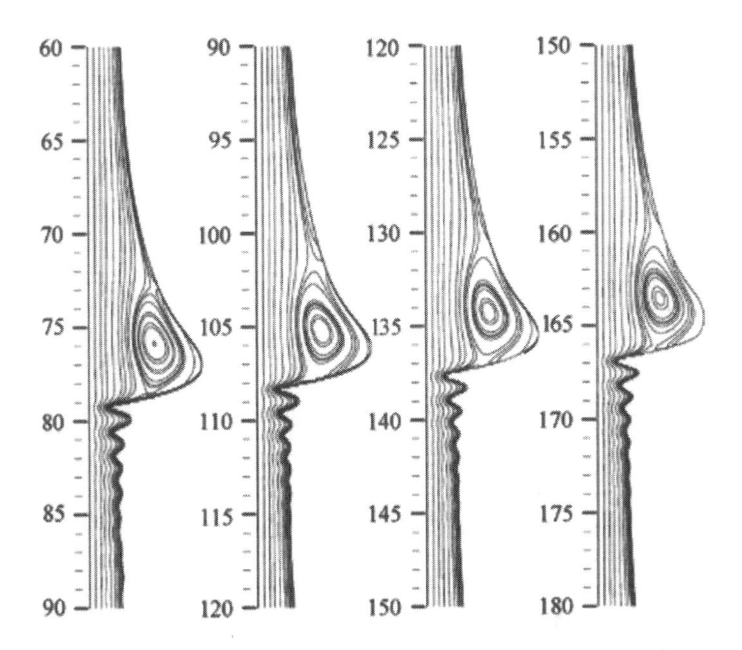

FIGURE 4.12 Streamline of the wavy film (Islam et al., 2009).

A numerical model is developed to study the heat and mass transfer coefficients of a double-effect LiBr–H_2O spiral tubular absorber (Hafsia et al., 2015a). The findings showed that the absorption is more in droplet mode (77%) than in falling film mode (26%). The lower portion of the absorber acted as a heat exchanger due to reduced mass transfer and increased heat transfer coefficient. This model is later extended to study the effects of surface tension on the three modes of a falling film by considering the wettability factor (Hafsia et al., 2015b). The outcomes revealed that the capillarity effect has a greater influence on droplet interfacial pressure than on falling film interfacial pressure. Because the capillarity effect is more pronounced in droplet mode, it enhances the global absorption, verified by experimental results.

Numerical simulation using CFD-Fluent is carried out for LiBr–H_2O falling film absorber to investigate the effect of Reynolds number on the heat and mass transfer coefficients (Zhang et al., 2015). The results indicated that the local mass transfer coefficient increases as the Reynolds number increases, while the local heat transfer coefficient decreases with an increase in the Reynolds number. The local mass transfer coefficient approaches its maximum value at the inlet, then rapidly decreases along the length of the plate and reaches a constant value.

The effect of variable properties on the absorption process of LiBr–H_2O is numerically simulated using the CFD-Fluent package by Bo et al. (2010). The simulation incorporated the real convective coolant boundary conditions, with the constant coolant heat transfer coefficient, but the coolant temperature varying linearly along the flow direction. The results indicated that, due to the property variations, the absorption capacity is increased by 6.5% compared to the constant property case, as reported in Fig. 4.13.

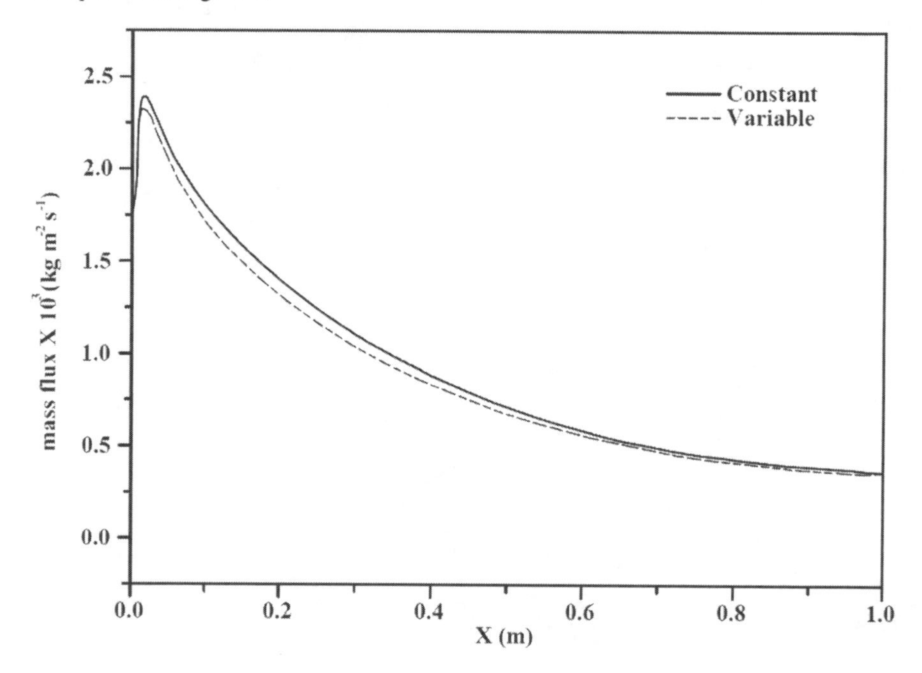

FIGURE 4.13 Mass flux along the film (Bo et al., 2010).

The effect of inlet sub-cooling on laminar wave-free film is numerically analysed by dividing the total absorption into two processes: Absorption due to sub-cooling of solution and absorption due to coolant wall (Chen and Christensen, 2000). The entire film is divided into small intervals to accomplish mixing in the solution. It is found that the absorption flux is greater with sub-cooling than without, but the absorption rate decreases with sub-cooling along the flow direction.

The effect of film waves and non-absorbable gases in vapour on mass transfer is studied numerically by Yang and Jou (1998). The findings revealed that mass transfer is enhanced due to waves but reduces in the presence of non-absorbable gases. Nusselt and Sherwood number correlations are developed in terms of Reynolds number and percentage of air. Similar studies of the 1-D model are developed for LiBr–H_2O laminar flow by Medrano et al. (2003). It is shown that a 61% reduction in absorption rate is due to an increase in air inlet concentration from 0 to 20%, as demonstrated in Fig. 4.14. High purge velocities are required to remove non-absorbable gases.

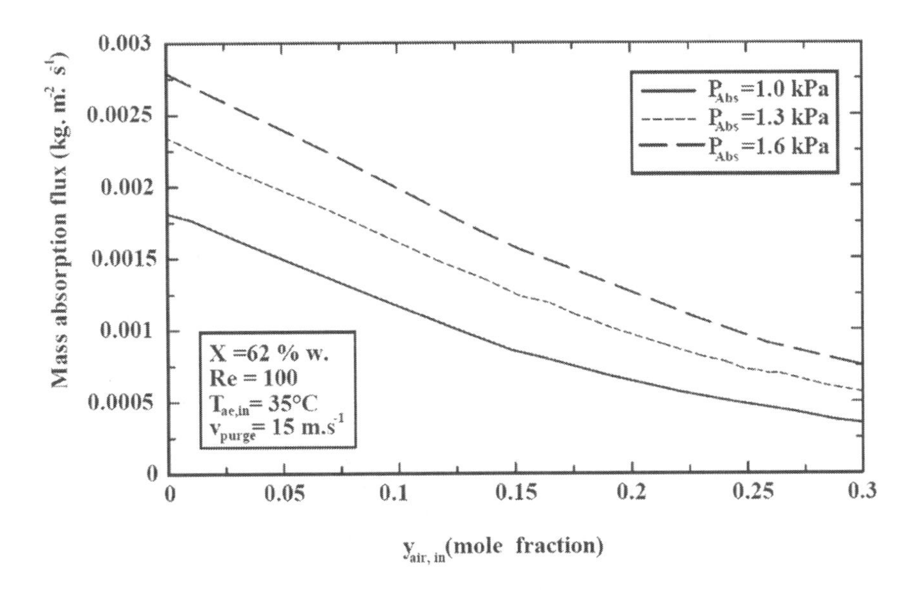

FIGURE 4.14 Effect of air concentration on absorption flux (Medrano et al., 2003).

A numerical model is developed to study the absorption phenomenon in the presence of nanoparticles (Yang et al., 2014). This model considered the effects of variation in film thickness, convection in film, and changes in ammonia–water nanofluid properties with a variation in ammonia concentration. The study also investigated the influence of the nanofluid's thermal conductivity, viscosity, mass transfer coefficient, and flow resistance on the absorption phenomenon. The model is validated with their own experimental results, with less than 20% relative errors.

Generally, falling film mode encounters wettability issues. To overcome this problem, researchers proposed the idea of using porous media to increase wettability in falling film. Numerical modelling of falling film (aqueous LiCl) absorption in porous media is carried out by Yang and Jou (1995). Results demonstrated that the use of porous media leads to an increase in both wettability and absorption rate. Also, an increase in absorption rate is observed with increasing conductivity ratio, solution flow rate, and absorber pressure increases. The optimum values for porosity and conductivity ratio are 0.9 and 10, respectively, for the range of operating conditions of 30–150 Re, 7–13 mm of Hg absorber pressure, 0.5–0.95 porosity, and 0.1–500 conductivity ratio. A correlation of Nusselt and Sherwood numbers is provided in terms of normalized pressure, porosity, and conductivity ratio.

Mathematical modelling of NH_3–H_2O falling film over a tube in the presence of an externally downward magnetic field is carried out to improve the absorption process (Xiaofeng et al., 2007). The study's findings revealed that the magnetic field positively impacts the absorption performance, and the absorption rate is observed to increase with the magnetic field intensity. The optimum magnetic field intensity is determined to be in the range of 0–3T for enhancing the absorption process. Notably, at a 3T magnetic field intensity of 3T, there is a 1.3% increase in outlet concentration, a 5.9% increase in absorbability, and a 4.73% increase in COP.

Cola et al. (2017) conducted a study to design a compact finned-plate absorber, which involved numerical simulations and experiments in optimizing fin shape. The finned-plate absorber is designed in such a way that it always ensures droplet flow, and a 3D printing technique is used to manufacture the fins. An analytical method based on a variational approach is developed to analyse the shape of fins, including circular, elliptic, rectangular, rhomboidal 60°, 90°, and 120°. Among these fin structures, rhomboidal 120° has the smallest drop size and is the easiest to manufacture. By using a 1 mm orifice with a rhomboidal 120° fin, a drop size of 2 mm could be achieved without droplet coalescence, resulting in a higher absorption rate.

Though many numerical studies on vertical falling film have existed in the literature, the effect of an inter-diffusion term is still unclear. To access the effect of inter-diffusion term and secondary effects like Dufour (heat flux due to concentration gradient) and Soret (mass flux due to temperature gradient) on vertical falling film, a CFD analysis is carried out by Hosseinnia et al. (2017b). This model is developed by considering constant parabolic velocity profile, laminar, wave-free film, variable heat of absorption, and variable thermophysical properties of $LiBr–H_2O$. The results showed that including an inter-diffusion term in the energy equation led to an 8.6% increase in the average heat flux and less than a 2% decrease in the average mass flux. Moreover, considering inter-diffusion, Dufour and Soret terms in the governing equations resulted in a 9.5% and 4% increment in the average heat and mass fluxes, respectively.

A CFD model is used to simulate the absorption phenomenon in a vertical swing wall absorber for $LiBr–H_2O$ (Gao et al., 2018). The wall swing caused the film to be affected by gravity and inertial forces, leading to variations in the surface characteristics. The model is analysed for a swing period of 4 s and a 10° swing amplitude. The result revealed five distinct regions under wall swing: Surface wave, monochromatic wave, quasi-sine wave at 0°, solitary wave at 4°, merge wave at 8°, and fall-off region at 10°. Additionally, experimental studies are conducted for TFE-TEGDME, with flow rates ranging from 0.2–0.5 kg min^{-1}, swing amplitudes ranging from 5–15°, and swing periods ranging from 4–12 s. The study divulged that the flow rate, swing period, and amplitude greatly affect the absorption rate. The maximum heat and mass flux occurred at a swing angle of 10° and a swing period of 8 s, while the least heat and mass flux occurred at a swing angle of 15° and swing periods of 4 s and 8 s, respectively.

All the studies mentioned above are related to numerical analysis. The experimental and enhancement studies are reported in Table 4.2.

TABLE 4.2
Experimental Studies on Vertical Surface Falling Film Absorber

References	Working Pair	Absorber Details	Operating Conditions	Results or Major Findings
Medrano et al. (2002); Takamatsu et al. (2003); Bourouis et al. (2005)	$LiBr–H_2O$ (Medrano et al., 2002) and ($LiBr + LiI + LiNO_3 + LiCl$) – H_2O (Bourouis et al., 2005)	Two stainless concentric tubes	L: 1.5 m Inner tube ID: 22.1 mm P: 1–2.2 kPa $X_{s,in}$: 57.9–64.2 wt% Re_s: 50–300 $T_{c,in}$: 30–45 °C Re_c: 6000	The effect of absorber pressure, solution concentration, solution flow rate, and cooling water temperature on mass transfer is discussed. No notable variation in absorption rate is observed between inner wall falling film and outer wall falling film. Similar studies are conducted by Takamatsu et al. (2003). Multi-component salt solution led to a 50% increase in absorption rate at 64.2% concentration compared to 57.9% (Bourouis et al., 2005).
Miller and Keyhani (2001)	$LiBr–H_2O$	Smooth tube	OD: 0.01905 m L: 1.524 m P: 1.3 kPa X_s: 0.6 Re_f: 290 T_c: 35 °C Re_c: 6000	Local heat and mass transfer rates are linear along the absorber length, resulting in an almost constant bulk concentration gradient. Nusselt and Sherwood's correlations are developed.

(*Continued*)

TABLE 4.2 *(Continued)*
Experimental Studies on Vertical Surface Falling Film Absorber

References	Working Pair	Absorber Details	Operating Conditions	Results or Major Findings
Li et al. (2018)	LiCl–H_2O and LiBr–H_2O	Two concentric tubes	Inner tube: copper, 1500 mm length, 15.88 mm diameter Outer tube: plexiglass, 250 mm ID P: 2–3 kPa $X_{s,in}$: 33–35.2% $T_{s,in}$: 45 °C Re_s: 275–483 $T_{c,in}$: 24–36 °C Re_c: 2500	Parametric studies are carried out. A numerical model is developed and validated with experimental data with an absolute error of 15%. A 19.7% higher absorption rate is achieved with a LiCl solution than with a LiBr solution. Nusselt and Sherwood number correlations are proposed with a deviation of 13%.
Oliva et al. (2016)	LiBr–H_2O	Vertical tube	ID: 0.018 m OD: 0.022 m L: 1.5 m P: 850–1260 kPa $X_{s,in}$: 57–61% $T_{s,in}$: 44.8–52 °C Re_s: 89–225 $T_{c,in}$: 30–42 °C	Mist flow (the presence of microdroplets of LiBr in water vapour) from the generator is identified, and it is increased with an increase in Reynolds number. Mist-free flow occurs when the Reynolds number is below 150. Observations are validated with a numerical model that accounts for the mist flow effects, smooth flow, and film wave characteristics.
Hao et al. (2014)	LiBr–H_2O	Bare brass tube and circumferentially coated brass tube	L: 1500 mm OD: 19 mm ID: 14 mm Coating: poly tetra fluoro ethylene (PTFE) Length of coating: 10 mm Coating thickness: 30 μm X_s: 55 wt% T_s: 120–140 °C m_s: 0.2–0.6 kg m^{-1} s^{-1} Re_c: 10,000	A parametric study reveals that the mass and heat transfer effects follow the same trend for both circumferentially coated and bare tubes. The circumferential coating enhances the absorption rate by improving mixing in the solution film thickness. The gap between two successive circumferential coatings also plays a role in absorption enhancement.

TABLE 4.2 *(Continued)*
Experimental Studies on Vertical Surface Falling Film Absorber

References	Working Pair	Absorber Details	Operating Conditions	Results or Major Findings
Kim et al. (1995)	LiBr–H_2O	Two concentric tubes with non-absorbable gases	Inner tube: steel, 38.1 mm OD, 1.83 m length Conc. of non-absorbable gases (air): less than 2% Corrosion inhibitor: 0.01 mol% LiOH P: 7.6–12 mm of Hg $X_{s,in}$: 57.7–60.5 mass % $T_{s,in}$: 35–45 °C Re_f: 15–150 $T_{c,in}$: 25–35 °C	When the air content is increased from 0.5 to 15% under given conditions, the Sherwood number decreases by 20%. The effect of non-absorbable gases remained constant if their concentration was below 2%. A dimensionless form of the mass transfer coefficient is developed.
Yoon et al. (2006)	LiBr–H_2O	Helical coil heat exchanger	P: 7 mm Hg $X_{s,in}$: 60 wt% $T_{s,in}$: 45–50 °C m_s: 50–210 kg h^{-1} Ref. vapour temp.: 7 °C $T_{c,in}$: 28–34 °C m_c: 300–900 kg h^{-1}	The helical absorber's heat and mass transfer coefficients are in the same order as the horizontal absorber. It is observed that the transfer coefficients increased significantly at low solution flow rates. However, this trend weakened as the flow rate increased, indicating the possibility of an optimal flow rate.
Kwon and Jeong (2004)	NH_3–H_2O	Helical coil heat exchanger	L: 600.6 mm D: 114.3 mm P: 0.17–1.93 bar Coil tube diameter: 12.7 mm No. of coil windings: 30 X_s: 3.13–30 $T_{s,in}$: 45–60 °C m_s: 4.43–79 g s^{-1} m^{-1} $T_{c,in}$: 30 °C	The effect of vapour flow direction is analysed. Parallel vapour flow resulted in better heat and mass transfer characteristics than in the counter flow direction. Correlations for Nusselt number are developed for both parallel and counterflow, and it is found that the interfacial shear effect is more pronounced in counterflow than in the parallel flow of vapour.

(Continued)

TABLE 4.2 *(Continued)*
Experimental Studies on Vertical Surface Falling Film Absorber

References	Working Pair	Absorber Details	Operating Conditions	Results or Major Findings
Yoon et al. (2005a)	$(LiBr + LiI + LiNO_3 + LiCl)-H_2O$	Concentric helical cylinder	P: 80.5 mm of Hg X_s: 0.58–0.64 T_s : 45–59 °C m_s: 0.01–0.04 kg m^{-2} s^{-1} T_c: 30–35 °C Velocity of coolant:1.43 m s^{-1}	The effect of solution flow rate on the absorption process is investigated. Absorber performance improves by 2–5% with a mixture of $(LiBr + LiI + LiNO_3 + LiCl)$ compared to using LiBr under the same operating conditions.
Kang et al. (1999)	NH_3–H_2O	Plate heat exchanger with offset strip fin	X_s: 0.05–0.15 T_s : 17–37.2 °C m_s: 4–10.15 g s^{-1} X_v: 0.647–0.797 m_v: 0.62–0.9 g s^{-1} T_v : 54.5–66.5 °C m_c: 98.83–121.25 g s^{-1}	Parametric studies are carried out. Due to inlet sub-cooling (where vapour temperature is greater than solution temperature), the rectification process occurred at the top of the absorber.
Miller and Perez-Blanco (1994)	LiBr–H_2O	Smooth, pin-fin, twisted, fluted and grooved tubes	OD: 0.019 m L: 1.53 m P: 8, 10 mm of Hg X_s: 60–64 wt% Re_s: 138.6–216.9 T_c: 35, 46 °C	Pin-fin (6.4 mm pitch) and grooved tubes increase the mass absorption rate by 225% and 175%, respectively, compared to a smooth tube. Increasing the cooling water flow rate with these advanced surfaces further enhanced the absorption rate.
Michel et al. (2017)	LiBr–H_2O	Vertical grooved plate	L: 220–420 mm P: 345–2635 Pa $X_{s,in}$: 57–60 wt% $T_{s,i}$: 20–40 °C Re_s: 15–350 m_c: 0 lph (adiabatic) 30–60 lph (isothermal)	Parametric studies are conducted, and the maximum absorption rate achieved is 7×10^{-3} kg m^{-2} s^{-1}. A numerical model is developed to validate the experimental results and determine the mass transfer effectiveness. The adiabatic mode has a greater mass transfer effectiveness than the isothermal mode, and this effectiveness increases with the Reynolds number.

TABLE 4.2 *(Continued)*
Experimental Studies on Vertical Surface Falling Film Absorber

References	Working Pair	Absorber Details	Operating Conditions	Results or Major Findings
Mortazavi et al. (2015)	$LiBr–H_2O$	Plate-and-frame with offset-strip fin	Fin: copper, 0.15 mm thickness, 6.35 mm lanced length, 6.35 mm fin height and 11.6 mm pitch. Water vapour pressure: 0.8–1.4 kPa $X_{s,in}$: 54–59 wt% $T_{s,in}$: 30–38 °C m_s: 0.44–1.22 kg min^{-1} m^{-1} $T_{c,in}$: 25–35 °C	The optimum gap between the fins is determined using CFD to achieve uniform film distribution and good wettability. The proposed configuration achieves a two-times higher absorption rate than a conventional falling film absorber.
Bor et al. (2015)	NH_3-H_2O	Vertical mini channel with annulus	L: 0.8 m D_h: 0.4 mm Eccentricity: 0.6 Tube size: 1.1 mm Annulus size: 1.6 mm ID, 2 mm OD X_s: 38.6% Re_s: 0–800 Vapour quality: 0–0.9	The heat transfer coefficient increases with the mass flow rate, heat flux, and inlet vapour quality. Pressure drop increases with the mass flow rate and vapour quality. As the pressure drop increases, the heat transfer coefficient initially increases, then remains constant before increasing again. This behaviour is due to the flow pattern changing from slug to annular.
Niu et al. (2010)	$NH_3–H_2O$	Vertical tube with an external magnetic field	L: 980 mm D: 140 mm P: 90 kPa Magnetic field intensity: 0–0.14 T $X_{s,in}$: 14.2–25.5% $T_{s,in}$: 30 °C m_s: 0.1 m^3 h^{-1} $T_{w,in}$: 18 °C m_w: 1 m^3 h^{-1}	A co-current magnetic field has a positive effect on the absorption process, whereas a counter-current magnetic field has a negative effect on the absorption rate. The effect of a magnetic field is effectively useful for lower inlet solution concentrations.

(Continued)

TABLE 4.2 *(Continued)*
Experimental Studies on Vertical Surface Falling Film Absorber

References	Working Pair	Absorber Details	Operating Conditions	Results or Major Findings
Hihara and Saito (1993)	LiBr–H_2O	Flat copper plate with surfactant	Surfactant: 2-Ethyl-1-hexanol Additive conc.: 60–100 ppm Absorber size: 50 mm wide, 300 mm long Inclination angle: 15–90 ° $X_{s,in}$: 0.573 $T_{s,in}$: 40 °C Re_{film}: 180–480 $T_{c,in}$: 30 °C m_c: 3 lpm	Transverse waves appeared as the flow rate increased, resulting in poor wetting and absorption. With additives, absorption is four to five times greater than without additives. The thickness and residence time of the falling film are reduced with the inclination angle. Marangoni instability is explained well.
Moller and Knoche (1996)	NH_3–H_2O	Plate heat exchanger with surfactants	Marlon PS (0.25, 0.4 wt%) Marlon A (0.25, 0.5 wt%) Dehydol LT 14 (0.5 wt%) Emulgin B 1 (0.5, 0.91 wt%) 1-octanol (50–500 ppm) P: 4.9 Kpa X_s: 0.25–0.91 T_s : 20 °C	Enhancement of absorption is possible with 1-octanol only.
Kim and Infante Ferreira (2009)	LiBr–H_2O	Bare copper plate and copper plate with copper wire screen. Also, with and without surfactant	Plate dimensions: 95 × 540 mm^2 Surfactant: 2-ethyl-1-hexanol with 100 ppm X_s: 50% Re_s: 40–110 Re_w: 1371–1403	Flow patterns on a bare tube are discussed, but flow patterns on a wire screen surface cannot be identified. 2E1H enhances the absorption rate by approximately two times that of bare tubes. Mass transfer enhancement of the wire screen is possible with LiBr–H_2O solution only.

TABLE 4.2 *(Continued)*
Experimental Studies on Vertical Surface Falling Film Absorber

References	Working Pair	Absorber Details	Operating Conditions	Results or Major Findings
Nordgren and Setterwall (1996)	Water–glycerol	Vertical falling film column with octanol	Surfactant conc.: 0–950 ppm Re: 0–250 X_s: .54 T_s: 25 °C	The effect of surfactants on film waviness is studied. The surface tension of the solution decreases sharply till 300 ppm of octanol.
Kim et al. (1996)	LiBr–H_2O	Concentric tubes with 2E1H	Additive conc.: 1–100 ppm P: 7.6 mm of Hg X_s: 0.6 T_s : 40 °C Re_f: 60 T_c : 30 °C Re_c: 10,000	Interfacial turbulence is initiated at a concentration of 3–6 ppm and reaches a maximum of 30 ppm.
Cheng et al. (2004)	LiBr–H_2O	Concentric tubes with 2E1H and 1-octanol	Additive conc.: 5–100 ppm P: 0.9 kPa X_s: 0.6 Re_f: 40–120 T_s : 40 °C T_c : 30 °C m_c: 1.2 kg min^{-1}	The degree of enhancement depends on the concentration of the additive and Reynolds number. The absorption process is enhanced in the presence of the additives.
Fu Lin and Shigang (2011)	LiBr–H_2O	Vertical falling film with 2E1H	Additive conc.: 90 ppm P: 1023 Pa X_s: 0.572–0.61 T_s: 35–45 °C T_c: 25.8–38.2 °C	The mass transfer coefficient is enhanced two times with additives compared to without additives under the same operating conditions.
Zhang et al. (2018)	LiBr–H_2O	Vertical falling film with Cu, Al_2O_3 and CNT with different particle sizes	Additive conc.: 0.01–0.1 wt% X_s: 0.57 T_s: 290 K V_s: 100–375 lph	The absorption phenomenon increases with the solution flow rate and nanoparticle concentration and decreases with the nanoparticle size. Cu nanoparticle has the best absorption enhancement.

REFERENCES

Álvarez, M.E., Bourouis, M., 2018. Experimental characterisation of heat and mass transfer in a horizontal tube falling film absorber using aqueous (lithium, potassium, sodium) nitrate solution as a working pair. Energy 148, 876–887. https://doi.org/10.1016/j.energy.2018.01.052

Aminyavari, M., Aprile, M., Toppi, T., Garone, S., Motta, M., 2017. A detailed study on simultaneous heat and mass transfer in an in-tube vertical falling film absorber. Int. J. Refrig. 80, 37–51. https://doi.org/10.1016/j.ijrefrig.2017.04.029

Bo, S., Ma, X., Lan, Z., Chen, J., Chen, H., 2010. Numerical simulation on the falling film absorption process in a counterflow absorber. Chem. Eng. J. 156, 607–612. https://doi.org/10.1016/j.cej.2009.04.066

Bor, D.M. Van De, Vasilescu, C., Ferreira, C.I., 2015. Experimental investigation of heat transfer and pressure drop characteristics of ammonia – water in a mini-channel annulus. Exp. Therm. FLUID Sci. 61, 177–186. https://doi.org/10.1016/j.expthermflusci.2014.10.027

Bourouis, M., Vallès, M., Medrano, M., Coronas, A., 2005. Absorption of water vapour in the falling film of water-(LiBr + LiI + LiNO3+ LiCl) in a vertical tube at air-cooling thermal conditions. Int. J. Therm. Sci. 44, 491–498. https://doi.org/10.1016/j.ijthermalsci.2004.11.009

Bredow, D., Jain, P., Wohlfeil, A., Ziegler, F., 2008. Heat and mass transfer characteristics of a horizontal tube absorber in a semi-commercial absorption chiller. Int. J. Refrig. 31, 1273–1281. https://doi.org/10.1016/j.ijrefrig.2008.01.016

Chen, W., Christensen, R.N., 2000. Inlet subcooling effect on heat and mass transfer characteristics in a laminar film flow. Int. J. Heat Mass Transf. 43, 167–177.

Cheng, W.L., Houda, K., Chen, Z.S., Akisawa, A., Hu, P., Kashiwagi, T., 2004. Heat transfer enhancement by additive in vertical falling film absorption of H2O/LiBr. Appl. Therm. Eng. 24, 281–298. https://doi.org/10.1016/j.applthermaleng.2003.08.013

Cola, F., Romagnoli, A., Heng Kiat, J.H., 2017. Optimisation of a compact falling-droplet absorber for cooling power generation. Energy Procedia 143, 354–360. https://doi.org/10.1016/j.egypro.2017.12.696

Cui, X., Shi, J., Tan, C., Xu, Z. (2009). Investigation of plate falling film absorber with film-inverting configuration. Journal of Heat Transfer 131(072001). https://doi.org/10.1115/1.3089550

Deng, S.M., Ma, W.B., 1999. Experimental studies on the characteristics of an absorber using LiBr/H2O solution as working fluid. Int. J. Refrig. 22, 293–301. https://doi.org/10.1016/S0140-7007(98)00067-X

Fu Lin, S.J., Shigang, Z., 2011. Experimental study on vertical vapor absorption into LiBr solution with and without additive. Appl. Therm. Eng. 31, 2850–2854. https://doi.org/10.1016/j.applthermaleng.2011.05.010

Fujita, I., Hihara, E., 2005. Heat and mass transfer coefficients of falling-film absorption process. Int. J. Heat Mass Transf. 48, 2779–2786. https://doi.org/10.1016/j.ijheatmasstransfer.2004.11.028

Fujita, T., 1993. Falling liquid films in absorption machines. Int. J. Refrig. 16, 282–294. https://doi.org/10.1016/0140-7007(93)90081-I

Gao, H., He, M., Sun, W., Yan, Y., 2018. Surface wave characteristic of falling film in swing absorber and its influences on absorption performance. Appl. Therm. Eng. 129, 1508–1517. https://doi.org/10.1016/j.applthermaleng.2017.09.141

Giannetti, N., Rocchetti, A., Lubis, A., Saito, K., Yamaguchi, S., 2016. Entropy parameters for falling film absorber optimisation. Appl. Therm. Eng. 93, 750–762. https://doi.org/10.1016/j.applthermaleng.2015.10.049

Giannetti, N., Rocchetti, A., Saito, K., Yamaguchi, S., 2015. Irreversibility analysis of falling film absorption over a cooled horizontal tube. Int. J. Heat Mass Transf. 88, 755–765. https://doi.org/10.1016/j.ijheatmasstransfer.2015.05.022

Giannetti, N., Rocchetti, A., Yamaguchi, S., Saito, K., 2017. Analytical solution of film mass-transfer on a partially wetted absorber tube. Int. J. Therm. Sci. 118, 176–186. https://doi.org/10.1016/j.ijthermalsci.2017.05.002

Goel, N., Goswami, D.Y., 2007. Experimental verification of a new heat and mass transfer enhancement concept in a microchannel falling film absorber. J. Heat Transfer 129, 154. https://doi.org/10.1115/1.2402182

Goel, N., Goswami, D.Y., 2005a. A compact falling film absorber. J. Heat Transfer 127, 957–965. https://doi.org/10.1115/1.1929781

Goel, N., Goswami, D.Y., 2005b. Analysis of a counter-current vapor flow absorber. Int. J. Heat Mass Transf. 48, 1283–1292. https://doi.org/10.1016/j.ijheatmasstransfer.2004.10.009

Hafsia, N. Ben, Chaouachi, B., Gabsi, S., 2015a. A study of the coupled heat and mass transfer during absorption process in a spiral tubular absorber. Appl. Therm. Eng. 76, 37–46. https://doi.org/10.1016/j.applthermaleng.2014.10.079

Hafsia, N. Ben, Chaouachi, B., Gabsi, S., 2015b. Surface tension effects on the absorption process in a spiral tubular absorber working with LiBr-H2O couple. Int. J. Therm. Sci. 94, 79–89. https://doi.org/10.1016/j.ijthermalsci.2015.02.009

Hao, Z., Lan, Z., Wang, Q., Zhao, Y., Ma, X., 2014. Heat and mass transfer enhancement for falling film absorption with coated distribution tubes at high temperature. Exp. Therm. Fluid Sci. 53, 147–153. https://doi.org/10.1016/j.expthermflusci.2013.11.022

Harikrishnan, L., Maiya, M.P., Tiwari, S., 2011a. Investigations on heat and mass transfer characteristics of falling film horizontal tubular absorber. Int. J. Heat Mass Transf. 54, 2609–2617. https://doi.org/10.1016/j.ijheatmasstransfer.2011.01.024

Harikrishnan, L., Tiwari, S., Maiya, M.P., 2011b. Numerical study of heat and mass transfer characteristics on a falling film horizontal tubular absorber for R-134a-DMAC. Int. J. Therm. Sci. 50, 149–159. https://doi.org/10.1016/j.ijthermalsci.2010.10.004

Hihara, E., Saito, T., 1993. Effect of surfactant on falling film absorption. Int. J. Refrig. 16, 339–346. https://doi.org/10.1016/0140-7007(93)90006-T

Ho, S., Soo, S., Kap, S., 2001. Wave characteristics of falling liquid film on a vertical circular tube. Int. J. Refrig. 24, 500–509.

Hoffmann, L., Greiter, I., Wagner, A., Weiss, V., Alefeld, G., 1996. Experimental investigation of heat transfer in a horizontal tube falling film absorber with aqueous solutions of LiBr with and without surfactants. Int. J. Refrig. 19, 331–341. https://doi.org/10.1016/S0140-7007(96)00026-6

Hosseinnia, S.M., Naghashzadegan, M., Kouhikamali, R., 2017a. CFD simulation of water vapor absorption in laminar falling film solution of water-LiBr – Drop and jet modes. Appl. Therm. Eng. 115, 860–873. https://doi.org/10.1016/j.applthermaleng.2017.01.022

Hosseinnia, S.M., Naghashzadegan, M., Kouhikamali, R., 2017b. Numerical study of falling film absorption process in a vertical tube absorber including Soret and Dufour effects. Int. J. Therm. Sci. 114, 123–138. https://doi.org/10.1016/j.ijthermalsci.2016.11.006

Hu, X., Jacobi, A.M., 1996. The intertube falling film : Part 1–flow characteristics, mode transitions, and hysteresis. J. Heat Transfer 118, 616–625.

Islam, M.A., Miyara, A., Setoguchi, T., 2009. Numerical investigation of steam absorption in falling film of LiBr aqueous solution with solitary waves. Int. J. Refrig. 32, 1597–1603. https://doi.org/10.1016/j.ijrefrig.2009.06.007

Islam, M.R., 2008. Absorption process of a falling film on a tubular absorber: An experimental and numerical study. Appl. Therm. Eng. 28, 1386–1394. https://doi.org/10.1016/j.applthermaleng.2007.10.004

Islam, M.R., Wijeysundera, N.E., Ho, J.C., 2003. Performance study of a falling-film absorber with a film-inverting configuration. Int. J. Refrig. 26, 909–917. https://doi.org/10.1016/S0140-7007(03)00078-1

Islam Md., R., Wijeysundera, N.E., Ho, J.C., 2004. Simplified models for coupled heat and mass transfer in falling-film absorbers. Int. J. Heat Mass Transf. 47, 395–406. https://doi.org/10.1016/j.ijheatmasstransfer.2003.07.001

Jeong, S., Garimella, S., 2002. Falling-film and droplet mode heat and mass transfer in a horizontal tube LiBr/water absorber. Int. J. Heat Mass Transf. 45, 1445–1458. https://doi.org/10.1016/S0017-9310(01)00262-9

Kang, Y.T., Akisawa, A., Kashiwagi, T., 1999. Experimental correlation of combined heat and mass transfer for NH3-H2O falling film absorption. Int. J. Refrig. 22, 250–262.

Kang, Y.T., Kashiwagi, T., 2002. Heat transfer enhancement by Marangoni convection in the NH3-H2O absorption process. Int. J. Refrig. 25, 780–788. https://doi.org/10.1016/S0140-7007(01)00074-3

Karami, S., Farhanieh, B., 2011. Numerical modeling of incline plate LiBr absorber. Heat Mass Transf. und Stoffuebertragung 47, 259–267. https://doi.org/10.1007/s00231-010-0715-2

Keast, P., Muir, P.H., 1991. EPDECOL. ACM Trans. Math. Soft. 17(2), 153–166.

Killion, J.D., Garimella, S., 2004. Simulation of pendant droplets and falling films in horizontal tube absorbers. J. Heat Transfer 126, 1003–1013. https://doi.org/10.1115/1.1833364

Killion, J. D., Garimella, S., 2003. A review of experimental investigations of absorption of water vapor in liquid films falling over horizontal tubes. HVAC&R Res. 9, 37–41. https://doi.org/10.1080/10789669.2003.10391060

Killion, Jesse D., Garimella, S., 2003. Gravity-driven flow of liquid films and droplets in horizontal tube banks. Int. J. Refrig. 26, 516–526. https://doi.org/10.1016/S0140-7007(03)00009-4

Killion, J.D., Garimella, S., 2001. A critical review of models of coupled heat and mass transfer in falling-film absorption. Int. J. Refrig. 24, 755–797. https://doi.org/10.1016/S0140-7007(00)00086-4

Kim, D.S., Ferreira, C.A.I., 2010. Effectiveness of non-volatile falling film absorbers with solution and coolant in counterflow. Int. J. Refrig. 33, 79–87. https://doi.org/10.1016/j.ijrefrig.2009.08.010

Kim, D.S., Ferreira, C.A.I., 2009. Analytic modelling of a falling film absorber and experimental determination of transfer coefficients. Int. J. Heat Mass Transf. 52, 4757–4765. https://doi.org/10.1016/j.ijheatmasstransfer.2009.05.014

Kim, D.S., Infante Ferreira, C.A., 2009. Flow patterns and heat and mass transfer coefficients of low Reynolds number falling film flows on vertical plates: Effects of a wire screen and an additive. Int. J. Refrig. 32, 138–149. https://doi.org/10.1016/j.ijrefrig.2008.08.005

Kim, H., Jeong, J., Kang, Y.T., 2012a. Heat and mass transfer enhancement of binary nanofluids for H2O/LiBr falling film absorption process. Int. J. Refrig. 35, 645–651. https://doi.org/10.1016/j.ijrefrig.2011.11.018

Kim, H., Jeong, J., Kang, Y.T., 2012b. Heat and mass transfer enhancement for falling film absorption process by SiO2 binary nanofluids. Int. J. Refrig. 35, 645–651. https://doi.org/10.1016/j.ijrefrig.2011.11.018

Kim, J., Cho, K., 2005. Optimisation of the absorption performance of a vertical absorber, in: Proceedings of HT2005 2005 ASME Summer Heat Transfer Conference 2005 ASME Summer Heat Transfer Conference July 17-22, 2005, San Francisco, California, USA. pp. 1–6.

Kim, J.K., Park, C.W., Kang, Y.T., 2003. The effect of micro-scale surface treatment on heat and mass transfer performance for a falling film H2O/LiBr absorber. Int. J. Refrig. 26, 575–585. https://doi.org/10.1016/S0140-7007(02)00147-0

Kim, K., Berman, N., Chau, D.S., Wood, B., 1995. Absorption of water vapour into falling films of aqueous lithium bromide. Int. J. Refrig. 18, 486–494. https://doi.org/10.1016/0140-7007(95)93787-K

Kim, K.J., Berman, N.S., Wood, B.D., 1996. The interfacial turbulence in falling film absorption: Effects of additives. Int. J. Refrig. 19, 322–330. https://doi.org/10.1016/S0140-7007(96)00025-4

Kwon, K., Jeong, S., 2004. Effect of vapor flow on the falling-film heat and mass transfer of the ammonia/water absorber. Int. J. Refrig. 27, 955–964. https://doi.org/10.1016/j.ijrefrig.2004.06.009

Kyung, I.-S., Herold, K.E., 2002. Performance of horizontal smooth tube absorber with and without 2-ethyl-hexanol. J. Heat Transfer 124, 177. https://doi.org/10.1115/1.1418366

Kyung, I., Herold, K.E., Tae, Y., 2007. Experimental verification of H2O/LiBr absorber bundle performance with smooth horizontal tubes. Int. J. Refrig. 30, 582–590. https://doi.org/10.1016/j.ijrefrig.2006.11.005

Li, T., Yin, Y., Liang, Z., Zhang, X., 2018. Experimental study on heat and mass transfer performance of falling film absorption over a vertical tube using LiCl solution. Int. J. Refrig. 85, 109–119. https://doi.org/10.1016/j.ijrefrig.2017.09.015

Medrano, M., Bourouis, M., Coronas, A., 2002. Absorption of water vapour in the falling film of water – lithium bromide inside a vertical tube at air-cooling thermal conditions. Int. J. of Thermal Sci. 41, 891–898.

Medrano, M., Bourouis, M., Perez-Blanco, H., Coronas, A., 2003. A simple model for falling film absorption on vertical tubes in the presence of non-absorbables. Int. J. Refrig. 26, 108–116. https://doi.org/10.1016/S0140-7007(02)00015-4

Meyer, T., 2015. Analytical solution for combined heat and mass transfer in laminar falling film absorption with uniform film velocity-diabatic wall boundary. Int. J. Heat Mass Transf. 80, 802–811. https://doi.org/10.1016/j.ijheatmasstransfer.2014.09.049

Meyer, T., 2014. Analytical solution for combined heat and mass transfer in laminar falling film absorption with uniform film velocity-Isothermal and adiabatic wall. Int. J. Refrig. 48, 74–86. https://doi.org/10.1016/j.ijrefrig.2014.08.005

Michel, B., Le Pierrès, N., Stutz, B., 2017. Performances of grooved plates falling film absorber. Energy 138, 103–117. https://doi.org/10.1016/j.energy.2017.07.026

Miller, W. a, Perez-Blanco, H., 1994. Vertical-tube aqueous LiBr falling film absorption using advanced surfaces. Int. Absoroption Heat Pump Conf. 185–202.

Miller, W.A., Keyhani, M., 2001. The correlation of simultaneous heat and mass transfer experimental data for aqueous lithium bromide vertical falling film absorption. J. Sol. Energy Eng. 123, 30. https://doi.org/10.1115/1.1349550

Moller, R., Knoche, K.F., 1996. Surfactants with NH3-H20. Int J. Refrig 19, 317–321.

Mortazavi, M., Moghaddam, S., 2016. Laplace transform solution of conjugate heat and mass transfer in falling film absorption process. Int. J. Refrig. 66, 93–104. https://doi.org/10.1016/j.ijrefrig.2016.02.013

Mortazavi, M., Nasr Isfahani, R., Bigham, S., Moghaddam, S., 2015. Absorption characteristics of falling film LiBr (lithium bromide) solution over a finned structure. Energy 87, 270–278. https://doi.org/10.1016/j.energy.2015.04.074

Nagavarapu, A.K., Garimella, S., 2013. Falling-film absorption around microchannel tube banks. J. Heat Transfer 135, 122001. https://doi.org/10.1115/1.4024261

Nakoryakov, V.E., Grigoryeva, N.I., Bartashevich, M. V., 2011. Heat and mass transfer heat and mass transfer in the entrance region of the falling film: Absorption, desorption, condensation and evaporation. Int. J. Heat Mass Transf. 54, 4485–4490. https://doi.org/10.1016/j.ijheatmasstransfer.2011.06.032

Narváez-Romo, B., Chhay, M., Zavaleta-Aguilar, E.W., Simões-Moreira, J.R., 2017. A critical review of heat and mass transfer correlations for LiBr-H2O and NH3-H2O absorption refrigeration machines using falling liquid film technology. Appl. Therm. Eng. 123, 1079–1095. https://doi.org/10.1016/j.applthermaleng.2017.05.092

Niu, X.F., Du, K., Xiao, F., 2010. Experimental study on ammonia-water falling film absorption in external magnetic fields. Int. J. Refrig. 33, 686–694. https://doi.org/10.1016/j.ijrefrig.2009.11.014

Nordgren, M., Setterwall, F., 1996. An experimental study of the effects of surfactant on a falling liquid film. Int J. Refrig 19, 310–316. https://doi.org/10.1109/ACC.2006.1657367

Oliva, A., Castro, J., Farn, J., 2016. Numerical and experimental investigation of a vertical LiBr falling film absorber considering wave regimes and in presence of mist flow. Int. J. Therm. Sci. 109, 342–361. https://doi.org/10.1016/j.ijthermalsci.2016.05.029

Papaefthimiou, V.D., Koronaki, I.P., Karampinos, D.C., Rogdakis, E.D., 2012. A novel approach for modelling LiBr-H2O falling film absorption on cooled horizontal bundle of tubes. Int. J. Refrig. 35, 1115–1122. https://doi.org/10.1016/j.ijrefrig.2012.01.015

Park, C.W., Cho, H.C., Kang, Y.T., 2004. The effect of heat transfer additive and surface roughness of micro-scale hatched tubes on absorption performance. Int. J. Refrig. 27, 264–270. https://doi.org/10.1016/j.ijrefrig.2003.09.008

Park, C.W., Kim, S.S., Cho, H.C., Kang, Y.T., 2003. Experimental correlation of falling film absorption heat transfer in micro-scale hatched tubes. Int. J. Refrig. 26, 758–763. https://doi.org/10.1016/S0140-7007(03)00069-0

Perez-Blanco, Tsai, B.-B., 1998. Limits of mass transfer enhancement in lithium bromide-water absorbers by active techniques. Int. J. Heat Mass Transf. 41, 2409–2416.

Qiu, Q., Meng, C., Quan, S., Wang, W., 2017. 3-D simulation of flow behaviour and film distribution outside a horizontal tube. Int. J. Heat Mass Transf. 107, 1028–1034. https://doi.org/10.1016/j.ijheatmasstransfer.2016.11.009

Seo, T., Cho, E., 2004. Coupled heat and mass transfer in absorption of water vapor into LiBr-HzO solution flowing on finned inclined surfaces. KSME Int. J. 18, 1140–1149.

Soto Francés, V.M., Pinazo Ojer, J.M., 2003. Validation of a model for the absorption process of H2O(vap) by a LiBr(aq) in a horizontal tube bundle, using a multi-factorial analysis. Int. J. Heat Mass Transf. 46, 3299–3312. https://doi.org/10.1016/S0017-9310(03)00121-2

Subramaniam, V., Garimella, S., 2014. Numerical study of heat and mass transfer in lithium bromide-water falling films and droplets. Int. J. Refrig. 40, 211–226. https://doi.org/10.1016/j.ijrefrig.2013.07.025

Takamatsu, H., Yamashiro, H., Takata, N., Honda, H., 2003. Vapor absorption by LiBr aqueous solution in vertical smooth tubes. Int. J. Refrig. 26, 659–666. https://doi.org/10.1016/S0140-7007(03)00038-0

Triché, D., Bonnot, S., Perier-Muzet, M., Boudéhenn, F., Demasles, H., Caney, N., 2017. Experimental and numerical study of a falling film absorber in an ammonia-water absorption chiller. Int. J. Heat Mass Transf. 111, 374–385. https://doi.org/10.1016/j.ijheatmasstransfer.2017.04.008

Triché, D., Bonnot, S., Perier-Muzet, M., Boudéhenn, F., Demasles, H., Caney, N., 2016. Modeling and experimental study of an ammonia-water falling film absorber. Energy Procedia 91, 857–867. https://doi.org/10.1016/j.egypro.2016.06.252

Wang, X., Jacobi, A.M., 2012. A thermodynamic basis for predicting falling-film mode transitions, in: International Refrigeration and Air Conditioning Conference at Purdue. https://doi.org/10.1016/j.ijrefrig.2014.04.002

Wu, Y., 2016. Simultaneous heat and mass transfer in laminar falling film on the outside of a circular tube. Int. J. Heat Mass Transf. 93, 1089–1099. https://doi.org/10.1016/j.ijheatmasstransfer.2015.11.031

Xiaofeng, N., Kai, D., Shunxiang, D., 2007. Numerical analysis of falling film absorption with ammonia – water in magnetic field. Appl. Therm. Eng. 27, 2059–2065. https://doi.org/10.1016/j.applthermaleng.2006.12.001

Yang, L., Du, K., Niu, X., Zhang, Y., Li, Y., 2014. Numerical investigation of ammonia falling film absorption outside vertical tube with nanofluids. Int. J. Heat Mass Transf. 79, 241–250. https://doi.org/10.1016/j.ijheatmasstransfer.2014.08.016

Yang, L., Du, K., Niu, X.F., Cheng, B., Jiang, Y.F., 2011. Experimental study on enhancement of ammonia-water falling film absorption by adding nanoparticles. Int. J. Refrig. 34, 640–647. https://doi.org/10.1016/j.ijrefrig.2010.12.017

Yang, R., Jou, D., 1995. Heat and mass transfer of absorption process for the falling film flow inside a porous medium. Int. J. Heat Mass Transf. 38, 1121–1126. https://doi.org/10.1016/0017-9310(94)00253-R

Yang, R., Jou, T.-M., 1998. Non-absorbable gas effect on the wavy film absorption process. Int. J. Heat Mass Transf. 41, 3657–3668.

Yoon, J.-I., Kwon, O.-K., Moon, C.-G., Lee, H.-S., Bansal, P., 2005a. Heat and mass transfer characteristics of a helical absorber using LiBr and LiBr+LiI+LiNO3+LiCl solutions. Int. J. Heat Mass Transf. 48, 2102–2109. https://doi.org/10.1016/j.applthermaleng.2005.05.009

Yoon, J.-I., Phan, T.-T., Moon, C.-G., Bansal, P., 2005b. Numerical study on heat and mass transfer characteristic of plate absorber. Appl. Therm. Eng. https://doi.org/10.1016/j.applthermaleng.2005.01.004

Yoon, J.-I., Phan, T.T., Moon, C.-G., Lee, H.-S., Jeong, S.-K., 2008. Heat and mass transfer characteristics of a horizontal tube falling film absorber with small diameter tubes. Heat Mass Transf. 44, 437–444. https://doi.org/10.1007/s00231-007-0261-8

Yoon, J., Kwon, O., Bansal, P.K., Moon, C., Lee, H., 2006. Heat and mass transfer characteristics of a small helical absorber. Appl. Therm. Eng. 26, 186–192. https://doi.org/10.1016/j.applthermaleng.2005.05.009

Yoon, J.I., Kim, E., Choi, K.H., Seol, W.S., 2002. Heat transfer enhancement with a surfactant on horizontal bundle tubes of an absorber. Int. J. Heat Mass Transf. 45, 735–741.

Zhang, L., Fu, Z., Liu, Y., Jin, L., Zhang, Q., Hu, W., 2018. Experimental study on enhancement of falling film absorption process by adding various nanoparticles. Int. Commun. Heat Mass Transf. 92, 100–106. https://doi.org/10.1016/j.icheatmasstransfer.2018.02.011

Zhang, L., Wang, Y., Fu, Y., Xing, L., Jin, L., 2015. Numerical simulation of H2O/LiBr falling film absorption process. Energy Procedia 75, 3119–3126. https://doi.org/10.1016/j.egypro.2015.07.644

Zhao, C.Y., Ji, W.T., Jin, P.H., Zhong, Y.J., Tao, W.Q., 2018. Hydrodynamic behaviors of the falling film flow on a horizontal tube and construction of new film thickness correlation. Int. J. Heat Mass Transf. 119, 564–576. https://doi.org/10.1016/j.ijheatmasstransfer.2017.11.086

5 Bubble Absorber

In bubble-type absorbers, refrigerant vapour coming from the evaporator is bubbled through the weak solution, either co-current or counter currently from the bottom of the absorber by using a vapour distributor like a set of nozzles, as shown in Fig. 5.1. Refrigerant vapour bubbles grow from the nozzle, detach, and rise to the top, surrounded by weak solution until they disappear. As a result, the interfacial area between the vapour and solution increases, resulting in good mixing and wettability characteristics between the vapour and liquid phase. This configuration provides better heat and mass transfer coefficients. However, there is a large pressure drop in both the liquid and vapour phases. The orifice and hydrostatic pressure of the solution column leads to an unfavourable pressure gradient in the vapour phase, which can adversely affect the vapour absorption rate and evaporator temperature. As a result, the height of the bubble absorber is limited. Additionally, due to the bulk flow of liquid, the coolant-solution heat transfer area is smaller than other absorber configurations. A bubble absorber does not require a solution distributor. Instead, it does require a vapour distributor.

The absorption capacity as well as heat and mass transfer coefficients in bubble absorbers mainly depend on the bubble dynamics. Several factors affect bubble dynamics, such as the size and shape of the nozzle orifice, its orientation and angle, working fluid properties, and the absorber type, size, and construction.

Experimental correlations of the mean diameter of initial bubbles, average bubble size, specific interfacial area, and liquid mass transfer coefficient for various working fluids, including water, glycol, aqueous sodium sulphite solution, and methanol are determined by Akita and Yoshida (1974). These correlations are applicable for column diameters up to 60 cm, superficial gas velocities up to 1500 m h^{-1}, and gas holdup to 30%; however, they may not be suitable for rapid absorption cases. The authors also reported that surface tension, liquid and gas densities, and system properties do not affect bubble size. Away from the sparger, orifice diameter has an insignificant effect on bubble diameter in the bulk solution region. Subsequently, the bubble size and swarm characteristics are analyzed by Kumar et al. (1976).

The experimental determination of bubble diameters is conducted for fluids with low and high viscosity (Bhavaraju et al., 1978). The study revealed that, for moderate gas flow rates, the bubble size near the orifice increases with one-third of the power of the gas flow rate and one-tenth of the power of liquid viscosity. Good liquid mixing and high interfacial area have been achieved in low-viscosity fluids with reasonable gas flow rates. However, bubble break-up is not observed in high-viscosity fluids and is related to turbulence in the solution vessel rather than the gas in the orifice.

Experiments on the NH_3–H_2O bubble absorber are carried out to visualize the bubble behaviour and determine the bubble size (Kang et al., 2002a). Departing

DOI: 10.1201/9781003485193-5

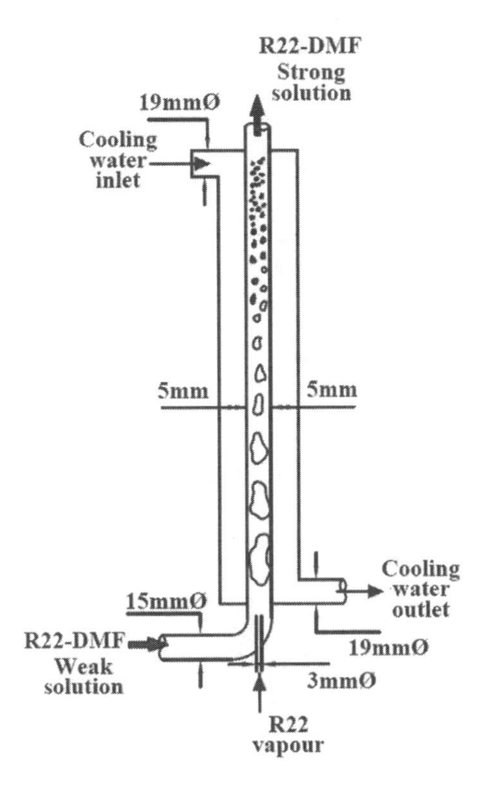

FIGURE 5.1 Bubble absorber (Sujatha et al., 1999).

bubbles tend to be spherical and hemispherical when the surface tension and inertia forces dominate, respectively. Increasing the orifice diameter resulted in a larger initial bubble diameter, which increased residence time and decreased the absorption rate. Also, the number of orifices has an insignificant effect on the initial bubble diameter. Finally, the authors developed a correlation for the initial bubble diameter from the experimental data with an error band of ±20%. However, Panda and Mani (2014) show that bubble diameter depends on the nozzle angle and the number of nozzles. Experimental correlations for bubble diameter are developed for H_2O–air in still, counter, and co-current flow using multi-tangential nozzles. The results showed that the initial diameter of the bubble increases with the nozzle angle.

Jiang et al. (2017a) found that a small orifice diameter can increase heat transfer but also lead to a larger pressure drop. In addition, the absorption height increases as the total flow area of the nozzle and vapour flow rate increases (Jiang et al., 2016), emphasizing the importance of selecting the appropriate nozzle size. Bubble visualization studies conducted by Jiang et al. (2016) and Kim et al. (2003a) demonstrated that the bubble absorber majorly consists of three flow regions, namely, churn, slug, and bubbly flow, as shown in Fig. 5.2. In general, for a given set of operating conditions, approximately 20% of the absorption height is churn flow, 65% is slug flow, and 15% is bubble flow (Ferreira et al., 1984).

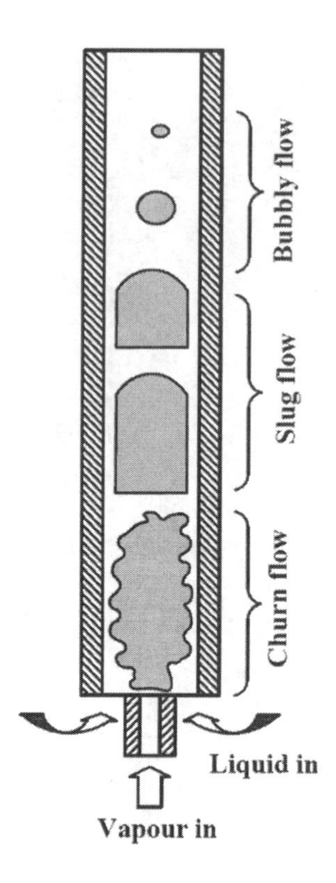

FIGURE 5.2 Flow patterns in vertical tubular absorber (Fernández-Seara et al., 2005).

Suresh and Mani (2012) conducted experiments to visualize the behaviour of the bubbles and determine the effect of gas flow rate and solution concentration on departing bubbles in still and flowing conditions. The experimental photograph of their study is shown in Fig. 5.3. They proposed an experimental correlation for the detachment of bubble diameter in R134a–DMF with an error of ±12%.

Analytical studies on the bubble behaviour in a sub-cooled liquid binary solution are carried out by Merrill (1997). The finite difference method solves the governing equations for the moving liquid–vapour interface. The predicted results agree with the semi-empirical relations available in the literature. The effects of counter-diffusion of the absorbent (water), initial bubble diameter, and absorber cooling rate on bubble collapse are investigated. The results showed that counter-diffusion of water extends the lifespan of bubbles, but this can be corrected by the convective mass transfer arising from interface collapse. The collapse time of bubbles increases linearly with the initial bubble size and the absorption rate increases with the square of the bubble radius. However, as the bubble size increases, the coalescence of bubbles can reduce the absorption efficiency by decreasing the effective mass transfer area. Rapid cooling of the solution reduces the bubble collapse time, but if the generated heat is removed too quickly, significant sub-cooling can occur, resulting in poor cycle efficiency (Merrill, 2000).

FIGURE 5.3 Experimental setup of glass bubble absorber system (Suresh and Mani, 2012).

A numerical model is developed to analyze the simultaneous heat and mass transfer process in a vertical tubular absorber working with an ammonia–water system (Ferreira et al., 1984). The model assumed that there are no heat and mass transfer resistances in the liquid and gas phases, that is, the concentration of ammonia is uniform in each phase, and the temperatures of both phases are equal and the same. The authors also extended their work by performing experimental analyses of 1 m length and internal diameters of 10 mm, 15.3 mm, and 20.5 mm with and without the heat of mixing removal. A correlation is developed to calculate the absorber length based on inlet conditions.

The absorption process in a vertical tubular bubble absorber is numerically modelled by Sujatha et al. (1997), which works with R22–DMF, R22–DMA, R22–DMETEG, and R22–NMP. The model is developed using the finite element method and Galerkin's technique. The authors have proposed a correlation for the mass transfer coefficient as a function of the Reynolds number, Schmidt number, and length-to-diameter ratio. The results showed that R22–DMA has better heat and mass transfer coefficients among all the working pairs.

The ammonia–water bubble absorption process is analyzed using a non-phenomenological theory, classic equilibrium phenomenological theory, and hydro-gas dynamics (Staicovici, 2000). This modelling approach contrasts empirical theories (e. g., two-film theory, penetration and renewal theory), experimental and Reynolds analogy correlations. The authors concluded that absorption is a mass phenomenon but not a surface phenomenon. For improving the absorption rate, the ideal point approaching (i.p.a) effect must occur, which can be achieved through the use of a bubble absorber. The authors also explain the advantages of a bubble absorber. This model can predict the reason for the Marangoni effect due to additives. Later, the phenomenological theory is extended to R134a–DMF (Suresh and Mani, 2010).

Mass transfer in a bubble absorber takes place in four stages: (1) mass transfer during bubble formation at an orifice; (2) mass transfer during the accelerating motion of the detached bubble; (3) mass transfer during the rise of the bubble with a constant velocity under unsteady state flow; and (4) mass transfer during the rise of the bubble with constant velocity under steady-state flow (Elperin and Fominykh, 2003). The authors have developed a numerical model using the similarity method to determine the simultaneous heat and mass transfer coefficients at each stage and its duration. The results showed that the residence time of the bubble increases with an increase in the initial bubble diameter and initial concentration.

Total absorption can be divided into three regions, namely churn, slug, and bubbly flow, as shown in Fig. 5.2. Fernández-Seara et al. (2005) developed a detailed numerical analysis for three different regimes that are taking place in vertical tubular absorbers working with ammonia–water. This model considered simultaneous heat and mass transfer in both liquid and vapour phases, and water is used as the coolant. The absorption process is very fast in churn and slug flow regimes but slow in bubbly flow. Heat transfer resistance is dominant in the vapour phase, and the authors conducted parametric studies to investigate this. The authors also observed that desorption of water from liquid to vapour occurred in churn flow and up to some length in slug flow. The same analysis is carried out with a vertical absorber using air as the coolant medium by Fernández-Seara et al. (2007).

The mathematical modelling of ammonia absorption is carried out to verify the experimental results (Lee et al., 2003). This model did not consider the counter-diffusion of absorbent into refrigerant bubbles. The results indicated that absorption height decreases with lower gas flow rates, higher overall heat transfer coefficients, lower cooling water temperatures, and lower input solution concentrations and temperatures. Furthermore, the absorption height is reduced by approximately 15 cm in the counter-flow mode compared to the co-current flow mode. A data reduction model based on the drift flux model and Chilton–Colburn analogy is developed for a slug flow absorber working with ammonia and water (Kim et al., 2003b). This model is able to predict local heat and mass transfer coefficients. The churn flow exhibited higher heat and mass transfer coefficients than the slug flow because of increased turbulence and mixing.

A plate-type bubble absorber is designed for R134a–DMAC (Mariappan et al., 2010). It is concluded that the plate-type bubble absorber provides better heat transfer than the shell and tube absorber, improving the overall heat transfer coefficient about 1.5–2 times greater than the shell and tube absorber. Furthermore, the size of the absorber is reduced accordingly.

Cerezo et al. (2010) conducted a study on the absorption of ammonia bubbles, compared to experimental results obtained from the Alfa Laval (NB51 model, type L) plate heat exchanger. The authors modified the conventional LMTD for subcooled-vapour and vapour–vapour solutions. The results showed that as the cooling water, solution flow rates, and pressure increase, the absorption performance also increases. However, as the concentration, cooling water and solution temperature increase, the absorption decreases. Later, the authors extended their modelling to NH_3–$LiNO_3$ and NH_3–NaSCN and concluded that NH_3–NaSCN has better absorption characteristics than the other two working pairs (Cerezo et al., 2011).

The effect of non-condensable gas (nitrogen) in ammonia vapour absorption into water is examined experimentally and numerically using a non-spherical bubble formation model (Terasaka et al., 2002). The study revealed that the bubble volume decreases as the gas flow rate or gas chamber volume decreases or the ammonia concentration in a bubble increases. It is observed that the interfacial mass transfer resistance remained almost constant, while the gas phase mass transfer resistance changed due to convection in the bubble. Moreover, the mass transfer resistance in the liquid phase is 100–1000 times greater than the other resistances.

Experiments are conducted to explain the mechanism of the Marangoni effect in bubble absorbers. Six additives, namely, 2-ethyl-1-hexagonal (2E1H), n-octanol (n-O), 2-octanol (2-O), 3-octanol (3-O), 4-octanol (4-O), n-decanol (n-D), 2-decanol (2-D), and 3-decanol (3-D) are added to NH_3–H_2O solution at concentrations ranging from 0 to 3000 ppm (Kang et al., 1999). The authors found that the solubility limit is between 500 and 3000 ppm, which is higher than LiBr–H_2O (70–400 ppm). The temperature gradient of surface tension and the concentration and temperature gradients of interfacial tensions are not the criteria of Marangoni convection in NH_3–H_2O. When ammonia concentration is increased without additives, surface tension decreases linearly. The study mentions the interfacial tension for all additives.

The addition of surfactants and nanofluids enhances heat transfer. In a study by Kim et al. (2007), the effect of binary nanofluids and surfactants is numerically analyzed for an ammonia–water bubble absorber. The considered surfactants are n-octanol, 2-octanol, and 2E1H, while the nanoparticles are copper (Cu), copper oxide (CuO), and alumina (Al_2O_3). It is observed that the absorption process is instant at the system inlet due to the addition of surfactants and nanofluids, which reduces the absorber's size. For a given set of conditions, adding 2E1H at 700 ppm reduced the length by 63%, and adding Cu at 1000 ppm reduced it by 54% compared to no additives to reach the solution concentration of 16.5%. Adding additives caused the mass transfer resistance to dominate over the heat transfer resistance.

Additional literature on this topic includes numerical and experimental investigations on LiBr–H_2O (Donnellan et al., 2014), experiments with plate heat exchange for NH_3–H_2O (Lee et al., 2002b), experiments with surface modifications of CNT for NH_3–H_2O (Xuehu et al., 2007; Ma et al., 2009), and experiments with nanoparticles of Cu, CuO, and Al_2O_3 for NH_3–H_2O (Kim et al., 2006a). The compilation of experimental studies and performance enhancement studies on bubble absorbers are presented in Table 5.1. Some of the experimental photographs are shown in Figs. 5.5 (Sujatha et al., 1999), 5.6 (Suresh and Mani, 2013), and 5.7 (Sanikommu et al., 2023).

TABLE 5.1
Experimental Studies on Bubble Absorber

References	Working Pair	Absorber Details	Operating Conditions	Remarks or Results or Major Findings
Kim et al. (2003a)	NH_3–H_2O	Pyrex glass tube for GAX cycle	P: 6 bar $X_{s,in}$: 2.5, 10.3, 21.6 wt% m_s: 1–3.5 kg h^{-1} m_g: 0.6, 0.9, 1.2 kg h^{-1} $T_{c,in}$: 20, 30, 40 °C m_c: 0.08–0.4 lpm	Flow pattern visualization is conducted. Most of the heat and mass transfer occurred in the churn flow rather than in the slug flow and bubbly flow. In slug flow, a smooth liquid film is observed.
Donnellan et al. (2014)	LiBr–H_2O	Glass column	L: 1 m D: 10 cm X_s: 46–56% T_s: 111–141 °C	Visualization of the single steam bubble in an aqueous LiBr solution is carried out. A higher initial bubble diameter resulted in a higher overall absorption rate.
Jiang et al. (2016)	R124–DMAC	Vertical glass tube with single and multi-orifices	L: 630 mm d_o: 1, 2, 2.2, and 2.8 mm P: 0.165 MPa $X_{s,in}$: 44 wt% $T_{s,in}$: 53 °C m_s: 4 lph m_v: 160–400 lph	Bubble flow patterns are visualized. As the total flow area of the nozzle and vapour flow rate increased, the absorption height also increased. For the same nozzle pressure difference, the use of multi-orifices is found to be ineffective.
Sanikommu et al. (2021)	Air–water	Vertical glass tube	L: 1000 mm ID: 85 mm OD: 89 mm m_w: 1–3.5 lpm m_a: 0.2–0.7 lpm	Bubble visualization studies are carried out with swirl entry of air in water. Bubbles' behaviour is studied in still and flowing water conditions. Bubble detachment diameter increases with air flow rate for all water conditions. The experimental photograph is shown in Fig. 5.4.

TABLE 5.1 *(Continued)*
Experimental Studies on Bubble Absorber

References	Working Pair	Absorber Details	Operating Conditions	Remarks or Results or Major Findings
Jiang et al. (2017a); Jiang et al. (2017b)	R124–DMAC	Copper vertical tube	P: 0.165 MPa d_o: 1, 2, 2.8 mm $X_{s,in}$: 35, 40, 45 wt% $T_{s,in}$: 36, 46, 55 °C m_s: 4, 6, 8 lph m_v: 80–400 lph T_c: 22, 32 °C m_c: 40 L h^{-1}	While a small orifice diameter can enhance heat transfer, it also causes a significant pressure drop, making the nozzle size a crucial factor in absorption. Nusselt and Sherwood number correlations are proposed.
Kang et al. (2002b)	NH_3–H_2O	Plate type	Absorber dimensions: 300 mm height, 80 mm length, and 53.4 mm width d_o: 3, 3.8, 5.5 mm P: 3 atm $X_{s,in}$: 0, 0.1, 0.2 $T_{s,in}$: 22.5 °C Inlet vapour velocity: 1.25–18.5 m s^{-1}	The absorption process is divided into two stages: Process I, which involves bubble growth, and process II, which pertains to bubble disappearance. Volumetric bubble diameter and Sherwood number correlations are developed in both stages. During the process I, the lower liquid concentrations increase the mass transfer coefficient. In process II, an increase in Galileo number, based on volumetric bubble diameter, corresponds to a higher mass transfer coefficient.
Sujatha et al. (1999)	HCFC22–DMF	Vertical tube	Inside tube dimensions: 18 mm ID and 25 mm OD L: 1800 mm No. of risers: six d_o: 3 mm P: 4.5–6 bar X_s: 0.52–0.65 m_s: 0.05–0.14 lps m_v: 0.002–0.005 lps T_c: 20–30 °C m_c: 0.25–0.4 lps	Heat and mass transfer coefficients and pressure drop values are calculated. Lower void fractions result in higher pressure drop. An increase in coolant flow rate led to higher pressure drops and improved mass transfer.

(Continued)

TABLE 5.1 *(Continued)*
Experimental Studies on Bubble Absorber

References	Working Pair	Absorber Details	Operating Conditions	Remarks or Results or Major Findings
Suresh and Mani (2011)	R134a–DMF	Vertical glass tube	P: 120–400 kPa $X_{s,in}$: 0.01–0.2 kg^{-1} $T_{s,\,in}$: 20–30 °C m_s: 50 lph m_v: 0.5–2.5 lpm m_c: 50 lph	The effect of gas flow rate, initial solution concentration, temperature, and pressure on the absorption rate is studied. Increased gas flow rate, solution initial concentration, and solution temperature correspond to higher heat and mass transfer coefficients. Conversely, a decrease in solution pressure also led to higher coefficients. Nusselt and Sherwood number correlations are provided.
Infante Ferreira (1985)	NH_3–$LiNO_3$ and –H_3-NaSCN	Vertical tube	ID: 10, 15.3, 20.5, 25.7 mm P: 120–370 kPa $X_{s,in}$: 0.35–0.45 kg kg^{-1} $T_{s,\,in}$: 20–40 °C m_s: 2–9.5 g s^{-1} T_v: 15–30 °C m_v: 0.02–0.32 g s^{-1} $T_{c,in}$: 20, 30, 40 °C	Temperature profiles are plotted by exclusively considering slug flow within the absorber.
Issa et al. (2002)	NH_3–H_2O	Rectangular cell	Dimensions: $12 \times 10 \times 1.5$cm P: 400 kPa X_s: 30.5, 42.4, 53.7, 59.4% T_s: 20 °C	When the initial concentration of a solution increases, the absorption rate initially decreases until it reaches 60%, after which absorption is ceased.
Suresh and Mani (2012)	R134a–DMF	Glass tube	ID: 33 mm OD: 37 mm L: 1000 mm P: 1.2–4 bar $X_{s,in}$: 0.01–0.2 kg kg^{-1} $T_{s,\,in}$: 20–30 °C V_s: 0.025–0.05 $m^3\,h^{-1}$ V_g: 0.03–0.15 $m^3\,h^{-1}$	Absorption and heat transfer rates do not vary much with the solution initial concentration of the solution. Parametric analysis is carried out.

TABLE 5.1 *(Continued)*
Experimental Studies on Bubble Absorber

References	Working Pair	Absorber Details	Operating Conditions	Remarks or Results or Major Findings
Lee et al. (2002b)	NH_3–H_2O	Plate type	Plate size: $0.112 \times 0.264 \times 0.03$ m^3 X_s: 0–30 wt% T_s: 20 °C m_s: 0–0.016 kg s^{-1} m_v: 0–1.6 × 10^4 m^3 s^{-1}	Increasing the solution flow rate greatly enhances heat transfer performance more than mass transfer enhancement. Increasing the gas flow rate enhances both heat and mass transfer performance. Nusselt and Sherwood number correlations are given.
Jung et al. (2014)	NH_3–H_2O	Plate type	P: 1150–1850 kPa X_s: 0.49–0.55 m_s: 0.09 kg s^{-1} T_c: 50 °C m_c: 0.05623 kg s^{-1}	Three plate heat exchangers with the same heat transfer area of 0.8 m^2 but varying in length, width, gap distance, aspect ratios (L/D and W/D) and the number of plates are evaluated. The aspect ratio influences the solution side heat transfer coefficient, as it increases with L/D. However, the influence of W/D on the heat transfer coefficient is much less.
Suresh and Mani (2013)	R134a-DMF	Brazed plate heat exchanger	m_s: 0.16–1.6 m^3 h^{-1} m_v: 0.03–0.09 m^3 h^{-1} T_c: 15–30 °C T_h: 67–95 °C m_h: 2.45 m^3 h^{-1} m_c: 2 m^3 h^{-1} (series) 1.25 m^3 h^{-1} (parallel)	An investigation is carried out to study the effect of operating parameters, such as absorber and generator temperatures and circulation ratio, on the overall heat transfer coefficient, volumetric mass transfer coefficient, and heat and mass transfer effectiveness. High generator temperatures and low circulation ratio improve heat and mass transfer effectiveness. Volumetric mass transfer coefficient correlation is proposed.
Oronel et al. (2013)	NH_3–$LiNO_3$ and NH_3– ($LiNO_3$ + H_2O)	Corrugated plate heat exchanger	P: 510 kPa $X_{s,in}$: 0.45–0.48 (binary mixture) 0.435 (ternary mixture) $T_{s,in}$: 45 °C m_s: 15–60 kg h^{-1} m_v: 0.6–2.2 kg h^{-1} $T_{c,in}$: 35–40 °C Water content in ternary mixture: 0.25 m_c: 130–333 kg h^{-1}	Mass absorption flux, heat transfer coefficient, mass transfer coefficient, and subcooling increase with an increase in solution flow rate for binary and ternary mixtures Lower viscosity and high affinity of ternary mixtures result in approximately 1.6 times higher mass transfer flux and 1.4 times higher heat transfer flux than binary mixtures. Nusselt and Sherwood number correlations are proposed.

(Continued)

TABLE 5.1 *(Continued)*
Experimental Studies on Bubble Absorber

References	Working Pair	Absorber Details	Operating Conditions	Remarks or Results or Major Findings
Cerezo et al. (2009)	NH_3–H_2O	Corrugated plate heat exchanger of model NB51, type L	Effective area of PHX: 0.1 m^2 d_o: 1.7 mm $X_{s,in}$: 0.3–0.38 $X_{s,out}$: 0.31–0.42 $T_{s,in}$: 35–55 °C $T_{s,out}$: 30–40 °C Re_s: 100–500 T_v: –8 to 0 °C Re_c: 200–700	The mass absorption flux ranged from 0.0025 to 0.0063 kg m^{-2} s^{-1}, while the mass transfer coefficient is between 0.001–0.002 m s^{-1}, and the solution heat transfer coefficient ranged from 2.7–5.4 kW m^{-2} K^{-1}. The amount of sub-cooling leaving the absorber is very low, indicating that almost complete ammonia absorption occurred.
Amaris et al. (2014a)	NH_3–$LiNO_3$	Vertical double pipe with a smooth inner surface and internally micro-finned surface	Micro fin length: 0.3 mm, 20° helix angle L: 1, 3 m D: 8, 9.5 mm P: 510 kPa $X_{s,in}$: 0.452 $T_{s,in}$: 45 °C m_s: 10–72 kg h^{-1} $T_{c,in}$: 35–40 °C m_c: 80–435 lph	When using micro finned tubes, absorption rate and heat transfer are improved by up to 1.7 and 1.55 times, respectively, when the solution mass flow rate is between 40 and 70 kg h^{-1}, without any pressure drop, as compared to using smooth tubes. The absorption mass flux increases with a decrease in the inner diameter and length of the micro-finned tube.
Cardenas and Narayanan (2011)	NH_3–H_2O	Large aspect ratio microchannel with smooth wall and stepped wall channels	Depth of microchannel: 600 µm Stepped channel trenches: 2 mm depth and 5mm width P: 6.2 bar X_s: 0.28 m_s: 10, 20, 35, and 55 g min^{-1} m_v: 1, 1.5, 2, 2.5, and 3 g min^{-1} $T_{c,in}$: 30, 40 and 58 °C	The absorption rates are higher for stepped wall geometry at lower coolant inlet temperatures of 30°C and 40°C, while the smooth wall channel exhibited higher mass transfer rates at a higher cooling temperature of 58°C.

TABLE 5.1 *(Continued)*
Experimental Studies on Bubble Absorber

References	Working Pair	Absorber Details	Operating Conditions	Remarks or Results or Major Findings
Jenks and Narayanan (2008)	NH_3–H_2O	Large aspect microchannel with six different geometries: three smooth micro channels of depth 150, 400, and 1500 μm, cross ribbed, 45° cross ribbed and streamwise finned	P: 2.5 and 4 bar $X_{s,in}$: 0–15% m_s: 10–30 g min^{-1} m_v: 1–3 g min^{-1}	A comparative study is conducted between six configurations based on its heat and mass transfer coefficients. For a constant vapour absorption rate, the absorber with a depth of 400 μm exhibited a higher overall heat transfer coefficient.
Kim et al. (2006b)	NH_3–H_2O	Bubble absorber with three kinds of surfactants: 2-Ethyl-1-hexanol (2E1H), n-octanol and 2-octanol	Surfactants concentration: 0–1000 ppm d_o: 2 mm Absorber: 20 mm width, 20 mm length, and 200 mm height P: 0.1 MPa $X_{s,in}$: 0–18.7% $T_{s,in}$: 20 °C	The shadow graphic method is used to visualize the bubbles. The addition of surfactants is found to enhance the absorption process. Among the three surfactants, 2E1H 700 ppm concentration exhibited the highest absorption ratio of 4.81 at an 18.7% solution. An increase in the initial solution concentration is observed to result in a higher absorption ratio. This indicates that adding surfactants can still increase the absorption rate even for a lower absorption potential. The barrier effect has a negligible effect on the absorption process, as the Marangoni effect is more dominant. An experimental correlation is provided for the effective absorption ratio (ratio of absorbed mass with a surfactant to without a surfactant).
Wu et al. (2013)	NH_3–H_2O	Bubble absorber with Fe_3O_4 nano ferrofluid and external magnetic fluid	Fe_3O_4 nano ferro fluid mass concentration: 0, 0.05, 0.1, 0.15, 0.2% External magnetic field intensity: 0, 80, 140, 208, 280 mT P: 250 kPa $X_{s,in}$: 0, 10, 20% $T_{s,in}$: 20 °C m_c: 0, 8 lpm	Fe_3O_4 nano ferrofluid with an external magnetic field enhances the absorption capacity up to some range. Increasing the intensity of the magnetic field results in a greater absorption capacity for a given nano-ferrofluid concentration. The combination of nano-ferrofluid and magnetic fields shows better performance in enhancing absorption than using them alone.

(Continued)

TABLE 5.1 ***(Continued)***
Experimental Studies on Bubble Absorber

References	Working Pair	Absorber Details	Operating Conditions	Remarks or Results or Major Findings
Amaris et al. (2014b)	NH_3–$LiNO_3$	Vertical double pipe with Carbon Nano Tube (CNT) and internally micro-finned surface	CNT concentration: 0–0.02 wt% P: 510 kPa $X_{s,in}$: 0.452 $T_{s,in}$: 45 °C m_s: 10–72 kg h^{-1} $T_{c,in}$: 35–40 °C m_c: 80–100 lph	The addition of CNT to the base working mixture improves the absorption flux by approximately 1.64 and 1.48 times that of a smooth tube at cooling water temperatures of 40 °C and 35 °C, respectively, without penalizing viscosity, pressure drop, and density. Combining CNT with advanced surface improved absorption flux for low solution flow rates. CNT showed similar absorption flux in the base fluid at a 0.01 and 0.02 wt% concentration.
Sanikommu et al. (2022); (Sanikommu et al., 2023)	R134a–DMF	Vertical copper absorber	Inner tube dimensions: ID: 50 mm OD: 55 mm Outer tube dimensions: ID: 70 mm OD: 75 mm m_s: 0.04–0.056 m^{-3} h^{-1} m_v: 0.005–0.011 m^{-3} h^{-1} $T_{c,in}$: 17–27 °C m_c: 0.08 m^{-3} h^{-1}	The effect of operating parameters on heat and mass transfer characteristics of bubble absorber with swirl entry of refrigerant has been studied. Overall heat transfer coefficient and volumetric mass transfer coefficients are increased with refrigerant and weak solution flow rates. Introducing swirl generator (SG) at the entry of refrigerant vapour enhances the solution heat transfer coefficient. The volumetric mass transfer coefficient is improved with SG.
Pang et al. (2012)	NH_3–H_2O	Bubble absorber with mono silver (Ag) solution	Absorber dimensions: Length: 20 mm Width: 20 mm Height: 200 mm d_o: 2 mm Nanoparticle diameter: 15 nm Nanoparticle concentration: 0–0.02 wt% P: 200kPa $X_{s,in}$: 0–20 wt% $T_{c,in}$: 15 °C	Adding 0.02 wt% Ag nanoparticles to the base fluid resulted in a 55% increase in absorption rate. This improvement in absorption can be due to an enhancement in heat transfer and the breaking of the gas bubble.

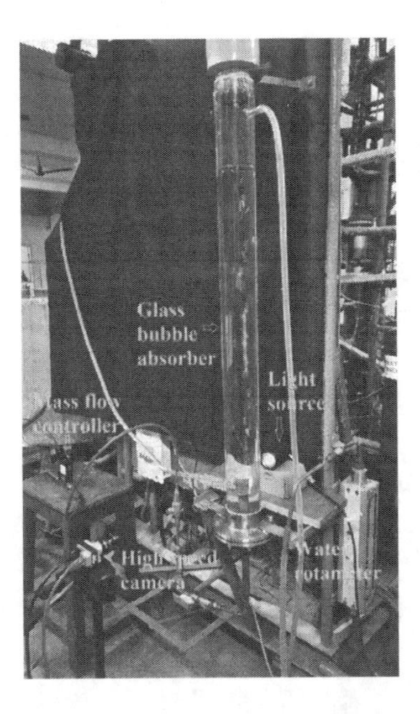

FIGURE 5.4 Photograph of the experimental setup with a swirl generator (Sanikommu et al., 2021).

FIGURE 5.5 Transfer tank operated VARS (Sujatha et al., 1999).

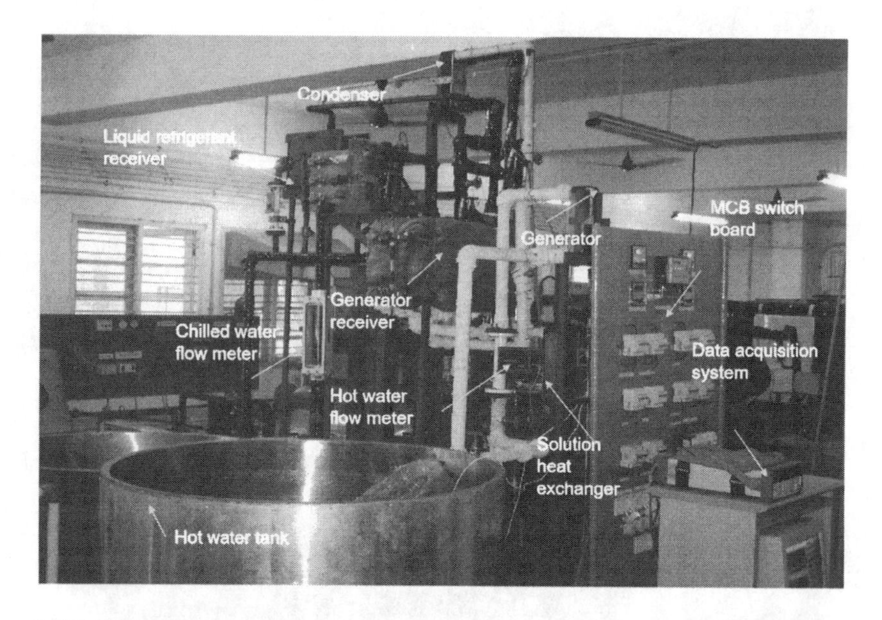

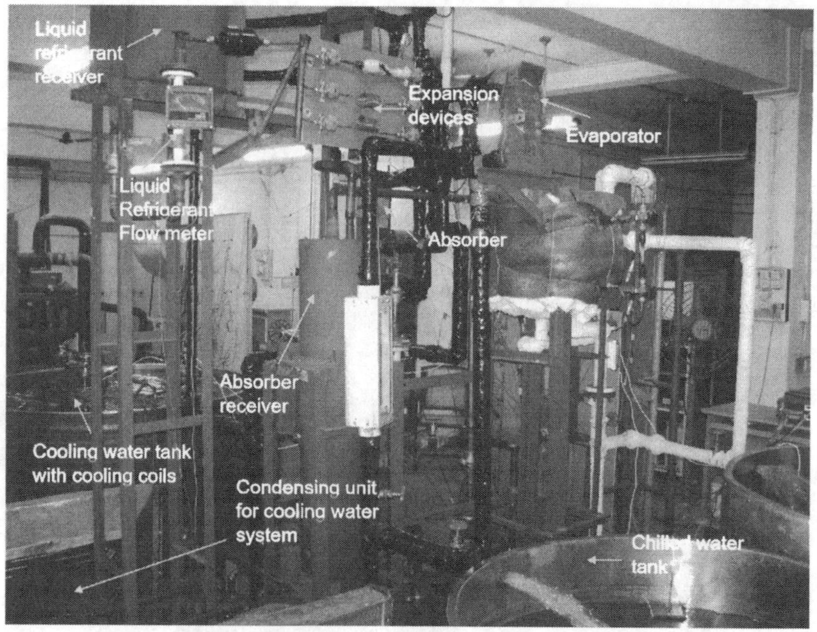

FIGURE 5.6 Experimental setup of vapour absorption refrigeration system (Suresh and Mani, 2013).

FIGURE 5.7 Photograph of the experimental setup with swirl generator (Sanikommu et al., 2023).

5.1 COMPARATIVE STUDIES ON FALLING FILM AND BUBBLE ABSORBER

In practical applications, falling film and bubble absorbers are commonly used. Therefore, these two configurations are compared in this section. The ammonia–water generator absorber heat exchanger (GAX) cycle is numerically analyzed to compare the falling film and bubble absorption modes using plate heat exchangers as an absorber in both modes (Kang et al., 2000). An offset strip fin is attached on the coolant side, as shown in Fig. 5.8. Mass and heat transfer

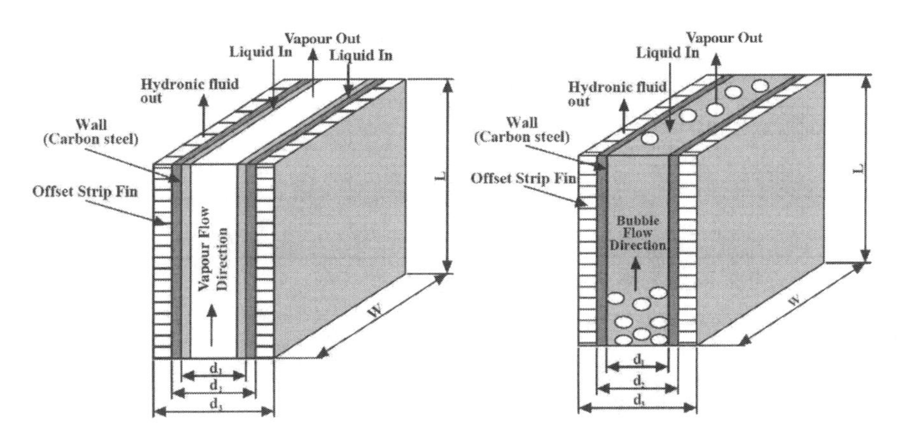

FIGURE 5.8 Falling film and bubble mode absorbers (Kang et al., 2000).

resistances are considered in the liquid and vapour regions for both modes. It is concluded that the absorption rate is better in the bubble mode than in the falling film mode, and a size reduction of approximately 49% is achieved with the bubble mode. The mass transfer resistance is dominant in the liquid flow, while the heat and mass transfer resistances are considerable in the vapour flow for the falling film mode. In contrast, the mass transfer resistance is dominant in the liquid phase, and the heat transfer resistance is dominant in the vapour phase for the bubble mode. The heat exchanger size is determined based on the heat transfer coefficients in the falling film mode and the mass transfer coefficients in the bubble mode.

Another numerical study compares the bubble and falling film modes for air-cooled vertical finned tubes working with ammonia–water (Castro et al., 2009). The model is validated using their experimental data and pre-existing numerical results, and the agreement is good. It is concluded that the bubble absorption mode outperformed the falling film mode, particularly at low solution flow rates.

Kang and Kim (2006) conducted an experimental comparison of LiBr–H_2O falling film and NH_3–H_2O bubble mode with the addition of surfactants (2E1H, n-octanol, and 2-octanol), nanofluids (Cu, CuO, and Al_2O_3), and surface roughness (0, 0.39, and 6.97 μm). For the falling film, the enhancement is 2.4, 3.8, and 4.5 times higher than the bare tube with 6.97 μm roughness, with surfactants and a combination of surface roughness and surfactants. In the bubble mode, the absorption ratio of 4.8 resulted in an 18.7% solution with 700 ppm 2E1H, and the absorption ratio of 3.2 is achieved with Cu nanoparticles. With the addition of surfactants and nanofluids, it is possible to enhance the absorption performance even at a lower mass transfer potential.

An experimental investigation is conducted on ammonia–water absorption using a plate-type absorber, comparing the performance of falling film and bubble absorption (Lee et al., 2002a). The results indicated that, despite the greater liberation of heat in bubble mode, it is more effective in terms of mass transfer. Furthermore, the effect of increasing the solution flow rate is more pronounced in bubble mode than in falling film. Ultimately, for low solution flow rate and high gas flow rate applications, bubble mode using a plate-type absorber is preferable.

A comparative experimental study is conducted to investigate the non-adiabatic absorption using falling film and bubble techniques (Helbing et al., 2000). The falling film is treated as a separated two-phase flow, while the bubble flow is considered a non-separated two-phase flow. Both falling film and bubble flow experiments are performed using a vertical tube. Additionally, bubble flow is tested using 30° and 60° corrugated plates. However, the corrugations had only a minor effect on heat transfer due to the less turbulent conditions in the rectangular channel. The results showed that bubble flow resulted in a higher liquid heat transfer coefficient and lower exergy loss than falling film. Therefore, bubble flow is found to be superior to falling film.

REFERENCES

Akita, K., Yoshida, F., 1974. Bubble size, interfacial area, and liquid-phase mass transfer coefficient in bubble columns. Ind. Eng. Chem. Process Des. Dev. 13, 84–91. https://doi.org/10.1021/i260049a016

Amaris, C., Bourouis, M., Vallès, M., 2014a. Effect of advanced surfaces on the ammonia absorption process with NH3/LiNO3in a tubular bubble absorber. Int. J. Heat Mass Transf. 72, 544–552. https://doi.org/10.1016/j.ijheatmasstransfer.2014.01.031

Amaris, C., Bourouis, M., Vallès, M., 2014b. Passive intensification of the ammonia absorption process with NH3/LiNO3using carbon nanotubes and advanced surfaces in a tubular bubble absorber. Energy 68, 519–528. https://doi.org/10.1016/j.energy.2014.02.039

Bhavaraju, S.M., Russell, T.W.F., Blanch, H.W., 1978. The design of gas sparged devices for viscous liquid systems. AIChE J. 24, 454–466. https://doi.org/10.1002/aic.690240310

Cardenas, R., Narayanan, V., 2011. Heat and mass transfer characteristics of a constrained thin-film ammonia-water bubble absorber. Int. J. Refrig. 34, 113–128. https://doi.org/10.1016/j.ijrefrig.2010.08.010

Castro, J., Oliet, C., Rodríguez, I., Oliva, A., 2009. Comparison of the performance of falling film and bubble absorbers for air-cooled absorption systems. Int. J. Therm. Sci. 48, 1355–1366. https://doi.org/10.1016/j.ijthermalsci.2008.11.021

Cerezo, J., Best, R., Bourouis, M., Coronas, A., 2010. Comparison of numerical and experimental performance criteria of an ammonia-water bubble absorber using plate heat exchangers. Int. J. Heat Mass Transf. 53, 3379–3386. https://doi.org/10.1016/j.ijheatmasstransfer.2010.02.031

Cerezo, J., Best, R., Romero, R.J., 2011. A study of a bubble absorber using a plate heat exchanger with NH3-H2O, NH3-LiNO3and NH3-NaSCN. Appl. Therm. Eng. 31, 1869–1876. https://doi.org/10.1016/j.applthermaleng.2011.02.032

Cerezo, J., Bourouis, M., Vallès, M., Coronas, A., Best, R., 2009. Experimental study of an ammonia-water bubble absorber using a plate heat exchanger for absorption refrigeration machines. Appl. Therm. Eng. 29, 1005–1011. https://doi.org/10.1016/j.applthermaleng.2008.05.012

Donnellan, P., Cronin, K., Lee, W., Duggan, S., Byrne, E., 2014. Absorption of steam bubbles in lithium bromide solution. Chem. Eng. Sci. 119, 10–21. https://doi.org/10.1016/j.ces.2014.07.060

Elperin, T., Fominykh, A., 2003. Four stages of the simultaneous mass and heat transfer during bubble formation and rise in a bubbly absorber. Chem. Eng. Sci. 58, 3555–3564. https://doi.org/10.1016/S0009-2509(03)00192-1

Fernández-Seara, J., Sieres, J., Rodríguez, C., Vázquez, M., 2005. Ammonia-water absorption in vertical tubular absorbers. Int. J. Therm. Sci. 44, 277–288. https://doi.org/10.1016/j.ijthermalsci.2004.09.001

Fernández-Seara, J., Uhía, F.J., Sieres, J., 2007. Analysis of an air cooled ammonia-water vertical tubular absorber. Int. J. Therm. Sci. 46, 93–103. https://doi.org/10.1016/j.ijthermalsci.2006.03.005

Ferreira, C.A.I., Keizer, C., Machielsen, C.H.M., 1984. Heat and mass transfer in vertical tubular bubble absorbers for ammonia-water absorption refrigeration systems. Int. J. Refrig. 7, 348–357. https://doi.org/10.1016/0140-7007(84)90004-5

Helbing, U., Würfel, R., Fratzscher, W., 2000. Comparative investigations of non-adiabatic absorption in film flow and bubble flow. Chem. Eng. Technol. 23, 1081–1085. https://doi.org/10.1002/1521-4125(200012)23:12<1081::AID-CEAT1081>3.0.CO;2-8

Infante Ferreira, C.A., 1985. Combined momentum, heat and mass transfer in vertical slug flow absorbers. Int. J. Refrig. 8, 326–334. https://doi.org/10.1016/0140-7007(85)90027-1

Issa, M., Ishida, K., Monde, M., 2002. Mass and heat transfers during absorption of ammonia into ammonia water mixture. Int. Commun. Heat Mass Transf. 29, 773–786. https://doi.org/10.1016/S0735-1933(02)00368-8

Jenks, J., Narayanan, V., 2008. Effect of channel geometry variations on the performance of a constrained microscale-film ammonia-water bubble absorber. J. Heat Transfer 130, 112402. https://doi.org/10.1115/1.2970065

Jiang, M., Xu, S., Wu, X., Hu, J., Wang, W., 2016. Visual experimental research on the effect of nozzle orifice structure on R124-DMAC absorption process in a vertical bubble tube. Int. J. Refrig. 68, 107–117. https://doi.org/10.1016/j.ijrefrig.2016.04.025

Jiang, M., Xu, S., Wu, X., 2017a. Experimental investigation for heat and mass transfer characteristics of R124-DMAC bubble absorption in a vertical tubular absorber. Int. J. Heat Mass Transf. 108, 2198–2210. https://doi.org/10.1016/j.ijheatmasstransfer.2017.01.082

Jiang, M., Xu, S., Wu, X., Wang, W., 2017b. Heat and mass transfer characteristics of R124-DMAC bubble absorption in a vertical tube absorber. Exp. Therm. Fluid Sci. 81, 466–474. https://doi.org/10.1016/j.expthermflusci.2016.09.008

Jung, C.W., An, S.S., Kang, Y.T., 2014. Thermal performance estimation of ammonia-water plate bubble absorbers for compression/absorption hybrid heat pump application. Energy 75, 371–378. https://doi.org/10.1016/j.energy.2014.07.086

Kang, Y., Akisawa, A., Kashiwagi, T., 1999. Visualization and model development of Marangoni convection in NH3–H2O system. Int. J. Refrig. 22, 640–649. https://doi.org/10.1016/S0140-7007(99)00019-5

Kang, Y.T., Akisawa, A., Kashiwagi, T., 2000. Analytical investigation of two different absorption modes: falling film and bubble types. Int. J. Refrig. 23, 430–443. https://doi.org/10.1016/S0140-7007(99)00075-4

Kang, Y.T., Kim, J.K., 2006. Comparisons of mechanical and chemical treatments and nano technologies for absorption applications. HVAC R Res. 12, 807–819. https://doi.org/10.1080/10789669.2006.10391209

Kang, Y.T., Nagano, T., Kashiwagi, T., 2002a. Visualization of bubble behavior and bubble diameter correlation for NH3- H2O bubble absorption. Int. J. Refrig. 25, 127–135.

Kang, Y.T., Nagano, T., Kashiwagi, T., 2002b. Mass transfer correlation of NH3-H2O bubble absorption. Int. J. Refrig. 25, 878–886. https://doi.org/10.1016/S0140-7007(01)00096-2

Kim, H.Y., Saha, B.B., Koyama, S., 2003a. Development of a slug flow absorber working with ammonia-water mixture: Part I - Flow characterization and experimental investigation. Int. J. Refrig. 26, 508–515. https://doi.org/10.1016/S0140-7007(03)00020-3

Kim, H.Y., Saha, B.B., Koyama, S., 2003b. Development of a slug flow absorber working with ammonia-water mixture: Part II - Data reduction model for local heat and mass transfer characterization. Int. J. Refrig. 26, 698–706. https://doi.org/10.1016/S0140-7007(03)00021-5

Kim, J.K., Jung, J.Y., Kang, Y.T., 2006a. The effect of nanoparticles on the bubble absorption performance in a binary nanofluid. Int. J. Refrig. 29, 22–29. https://doi.org/10.1016/j.ijrefrig.2005.08.006

Kim, J.K., Jung, J.Y., Kim, J.H., Kim, M.G., Kashiwagi, T., Kang, Y.T., 2006b. The effect of chemical surfactants on the absorption performance during NH3/H2O bubble absorption process. Int. J. Refrig. 29, 170–177. https://doi.org/10.1016/j.ijrefrig.2005.06.006

Kim, J.K., Akisawa, A., Kashiwagi, T., Kang, Y.T., 2007. Numerical design of ammonia bubble absorber applying binary nanofluids and surfactants. Int. J. Refrig. 30, 1086–1096. https://doi.org/10.1016/j.ijrefrig.2006.12.011

Kumar, A., Degaleesan, T.E., Laddha, G.S., Hoelscher, H.E., 1976. Bubble swarm characteristics in bubble columns. Can. J. Chem. Eng. 54, 503–508. https://doi.org/10.1002/cjce.5450540525

Lee, K.B., Chun, B.H., Lee, J.C., Hyun, J.C., Kim, S.H., 2002a. Comparison of heat and mass transfer in falling film and bubble absorbers of ammonia-water. Exp. Heat Transf. 15, 191–205. https://doi.org/10.1080/08916150290082621

Lee, K.B., Chun, B.H., Lee, J.C., Lee, C.H., Kim, S.H., 2002b. Experimental analysis of bubble mode in a plate-type absorber. Chem. Eng. Sci. 57, 1923–1929. https://doi.org/10.1016/S0009-2509(02)00089-1

Lee, J.C., Lee, K.B., Chun, B.H., Lee, C.H., Ha, J.J., Kim, S.H., 2003. A study on numerical simulations and experiments for mass transfer in bubble mode absorber of ammonia and water. Int. J. Refrig. 26, 551–558. https://doi.org/10.1016/S0140-7007(03)00002-1

Mariappan, V., Anand, R.B., Udayakumar, M., 2010. A simplified design procedure of R134A-DMAC plate type bubble absorber for vapour absorption refrigeration system, in: Proceedings of the International Conference on Frontiers in Automobile and Mechanical Engineering - 2010, FAME-2010. pp. 81–87. https://doi.org/10.1109/FAME.2010.5714804

Ma, X., Su, F., Chen, J., Bai, T., Han, Z., 2009. Enhancement of bubble absorption process using a CNTs-ammonia binary nanofluid. Int. Commun. Heat Mass Transf. 36, 657–660. https://doi.org/10.1016/j.icheatmasstransfer.2009.02.016

Merrill, T., 1997. Combined heat and mass transfer during bubble absorption in binary solutions. Int. J. Heat Mass Transf. 40, 589–603. https://doi.org/10.1016/0017-9310(96)00118-4

Merrill, T.L., 2000. Thermally controlled bubble collapse in binary solutions. Int. J. Heat Mass Transf. 43, 3287–3298. https://doi.org/10.1016/S0017-9310(99)00383-X

Oronel, C., Amaris, C., Bourouis, M., Vallès, M., 2013. Heat and mass transfer in a bubble plate absorber with NH3/LiNO3 and NH3/(LiNO3+ H2O) mixtures. Int. J. Therm. Sci. 63, 105–114. https://doi.org/10.1016/j.ijthermalsci.2012.07.007

Panda, S.K., Mani, A., 2014. Experimental study on bubble absorber with multiple tangential nozzles, in: 15th Int. Refrig. Air Cond. Conf. Purdue 1–10.

Pang, C., Wu, W., Sheng, W., Zhang, H., Kang, Y.T., 2012. Mass transfer enhancement by binary nanofluids (NH3/H2O + Ag nanoparticles) for bubble absorption process. Int. J. Refrig. 35, 2240–2247. https://doi.org/10.1016/j.ijrefrig.2012.08.006

Sanikommu, N.R., Mani, A., Tiwari, S., 2021. Bubble dynamics studies in an absorber with swirl entry of an absorption refrigeration system, in: 18th International Refrigeration and Air Conditioning Conference at Purdue, USA, May 24–28. pp. 1–9.

Sanikommu, N.R., Mani, A., Tiwari, S., 2022. Experimental studies on a bubble absorber with swirl entry of refrigerant vapour, in: 19th International Refrigeration and Air Conditioning Conference at Purdue, USA, July 10–14 2284, 1–10.

Sanikommu, N.R., Mani, A., Tiwari, S., 2023. Performance evaluation of a bubble absorber in a VARS with swirl entry of refrigerant vapour, in: 8th Therm. Fluids Eng. Conf. (TFEC-2023), Univ. Maryland, Coll. Park. MD, USA March, 26–29, 2023.

Staicovici, M.D., 2000. A phenomenological theory of polycomponent interactions in non-ideal mixtures. Application to NH3/H2O and other working pairs. Int. J. Refrig. 23, 153–167.

Sujatha, K.S., Mani, a., Murthy, S.S., 1997. Analysis of a bubble absorber working with R22 and five organic absorbents. Heat Mass Transf. 32, 255–259. https://doi.org/10.1007/s002310050119

Sujatha, K.S., Mani, A., Murthy, S.S., 1999. Experiments on a bubble absorber. Int. Commun. Heat Mass Transf. 26, 975–984. https://doi.org/10.1016/S0735-1933(99)00087-1

Suresh, M., Mani, A., 2010. Heat and mass transfer studies on R134a bubble absorber in R134a/DMF solution based on phenomenological theory. Int. J. Heat Mass Transf. 53, 2813–2825. https://doi.org/10.1016/j.ijheatmasstransfer.2010.02.016

Suresh, M., Mani, A., 2011. Evaluation of heat and mass transfer coefficients for R134a/DMF bubble absorber. J. Appl. Sci. 11, 1921–1928.

Suresh, M., Mani, A., 2012. Experimental studies on heat and mass transfer characteristics for R134a-DMF bubble absorber. Int. J. Refrig. 35, 1104–1114. https://doi.org/10.1016/j.ijrefrig.2012.01.011

Suresh, M., Mani, A., 2013. Heat and mass transfer studies on a compact bubble absorber in R134a-DMF solution based vapour absorption refrigeration system. Int. J. Refrig. 36, 1004–1014. https://doi.org/10.1016/j.ijrefrig.2012.10.033

Terasaka, K., Oka, J., Tsuge, H., 2002. Ammonia absorption from a bubble expanding at a submerged orifice into water. Chem. Eng. Sci. 57, 3757–3765. https://doi.org/10.1016/S0009-2509(02)00308-1

Wu, W.D., Liu, G., Chen, S.X., Zhang, H., 2013. Nanoferrofluid addition enhances ammonia/water bubble absorption in an external magnetic field. Energy Build. 57, 268–277. https://doi.org/10.1016/j.enbuild.2012.10.032

Xuehu, M., Su, F., Chen, J., Zhang, Y., 2007. Heat and mass transfer enhancement of the bubble absorption for a binary nanofluid. J. Mech. Sci. Technol. 21, 1813–1818. https://doi.org/10.1007/BF03177437

6 Comparison of Absorber Configurations

Table 6.1 discusses the important parameters that impact the overall performance of any absorber, including interfacial area, wettability, solution holdup, pressure drop, solution and vapour flow rates, solution concentration, and inlet temperature, as well as cooling water temperature and flow rate. Furthermore, Table 6.1 also covers the design complexity, manufacturing cost, and flexibility of various absorbers.

6.1 CONCLUSIONS

This chapter provides a comprehensive overview of the various types of absorbers used in VARS, including tray, packed bed, falling film, spray, bubble, and membrane absorbers, as well as the working fluids commonly used. The numerical, experimental, and performance enhancement studies of each absorber are discussed in detail, along with the operating conditions in the previous chapters. Most enhancement studies focused on falling film and bubble absorbers due to their extensive usage in practical applications, with surface modifications, the addition of surfactants, and nanoparticles being among the discussed methods. This chapter concludes by comparing all the absorbers based on their performance parameters.

Based on the findings presented in Table 6.1, it can be concluded that falling film and plate-type absorbers face several challenges, such as wettability issues, film break-up, non-uniform film thickness, and significant pressure drop. While falling film absorbers have better heat transfer characteristics, these are unsuitable for handling viscous fluids and have low mass transfer coefficients. Although wettability issues and film breakdown cannot be completely eliminated, the use of membranes can help maintain a uniform film thickness. However, despite offering high interfacial area, spray and packed bed absorbers suffer from effective heat rejection issues. For instance, external heat exchangers are required for heat rejection in spray absorbers, while packed bed absorbers require periodic replacement of packing materials. On the other hand, bubble absorbers have high interfacial area and good wettability characteristics and offer effective heat removal with external coolants. However, the complex bubble dynamics associated with bubble absorbers make modelling difficult. Recently, membrane absorbers have been developed for low-capacity applications, which offer constant solution film thickness and better interfacial area. Nonetheless, this absorber cannot operate at higher temperatures, and the strength of the membrane is low. It is important to note that all research on membrane absorbers has been conducted using conventional working fluids, and further research is required to determine their suitability with alternative working pairs.

DOI: 10.1201/9781003485193-6

TABLE 6.1
Comparison of Various Absorber Configurations

Characteristics	Tray Absorber	Packing bed Absorber	Falling Film Absorber	Spray Absorber	Bubble Absorber
The interfacial area between solution and vapour	Large interfacial area possible with bubble cap type plate	High interfacial area because of packing material	Low, due to film instabilities	High, because of spraying of weak solution	High due to the bubbling phenomenon, even at low solution flow rates.
Wettability problem	Serious, but it can be easily resolved when compared to falling film	It suffers a wettability problem if the minimum liquid flow rate is not satisfied	High, due to less interfacial area	It occurs only due to improper distribution of solution through the nozzle	No, because of the large interfacial area
Pressure drop	High	Low compared to tray absorber	Low compared to the tray and packed bed absorbers	High if the selection of solution distributor (nozzle) is inappropriate	Low compared to the tray and packed bed absorbers
Viscous working pair	Moderate in handling viscous liquids	Very difficult	Difficult	It depends on the nozzle selection	Easy to handle viscous fluids
Mass flow rates of solution and vapour	Suitable for high solution mass flow rate and low vapour mass flow rate	Suitable for low solution flow rates	Can handle high flow rates of solution and vapour	Can handle high flow rates of solution and vapour	Suitable for low solution flow rates.
Removal of heat from the absorber	Possible with some difficulties	Not possible	Possible	Effective heat rejection because of separate single-phase heat exchanger	Possible
Heat transfer rate	Higher than packed bed	Lower than all the configurations	Higher than tray, packed bed, and bubble mode	Higher than all the modes	Higher than tray absorber
Mass transfer rate	High compared to falling film but low compared to bubble and spray absorbers	High	Low	High	High
Efficiency	Moderate	Moderate	Same as that of the sieve tray absorber	High	More than falling film and tray absorbers
Design difficulty	Medium	High, due to difficulty associated with the selection of packing material	High because size increases as cooling capacity increases	High, because of difficulties in the selection of nozzle	Low
Flexibility	Quite flexible	Not so flexible	Quite flexible	Not so flexible	Quite flexible
Manufacturing cost	Inexpensive	Highly expensive	Highly expensive	Highly expensive	Moderately expensive

In summary, each absorber configuration has its own advantages and disadvantages, making it crucial to carefully select the appropriate configuration based on the specific application requirements. The bubble absorber may be preferred for low solution flow rate applications due to its higher heat and mass transfer coefficients, high interfacial area, and ease of heat removal. However, the literature review revealed a discrepancy between numerical and experimental results due to inadequate assumptions and limitations in numerical methods, indicating a need for improved modelling techniques to capture the absorption phenomenon accurately. Furthermore, the exploration of alternative working fluids that can enhance performance significantly, with lower circulation ratios and cut-off temperatures, should be pursued. Finally, research should focus on developing enhancement techniques to improve absorption performance.

Index

Note: *Italicized* page references refer to the figures and **bold** references refer to the tables.